More Simple Times at

Old Field Farm

By Suzy Lowry Geno

More Simple Times at Old Field Farm

More Simple Times at
Old Field Farm

Copyright © 2016 by Suzy Lowry Geno

All rights reserved.

Printed in the United States of America. No part of this book may be used or reproduced in any manner whatsoever without written permission except in the case of brief quotations embodied in critical articles and reviews.

Fifth Estate Publishing, Blountsville, AL 35031

Cover Designed by An Quigley

Printed on acid-free paper

Library of Congress Control No: 2016950358

ISBN: 9781936533855

Fifth Estate, 2016

Suzy Lowry Geno

More Simple Times at Old Field Farm

Suzy Lowry Geno

Table of Contents

1. **Shadow** — 9
2. **Professional Porch Sittin'** — 15
3. **The Roosters Ninety and Nine** — 21
4. **Tired but no retired** — 26
5. **Smells Take me Back** — 33
6. **Happy you happy** — 42
7. **Weird again** — 46
8. **Poor but rich is what some folks can't understand** — 50
9. **The Turkey Experiment** — 56
10. **I am an addict and it's great** — 62
11. **Being prepared in a different way** — 70
12. **Christmas memories from Old Field Farm** — 76
13. **Sometimes I wonder** — 81
14. **Guineas my latest love on the homestead** — 87
15. **You aren't here** — 94
16. **Wood heat warms your soul** — 101

17. Rabbits on the homestead — 108

18. High tech or low tech — 117

19. Wired or unwired — 123

20. Time doesn't matter to a pig — 126

21. Made in the USA — 132

22. The right to dry — 137

23. A Chance to laugh — 143

24. Finding God in the barnyard — 149

25. Work with your hands — 156

26. Not a Lone Ranger — 162

27. Thanks page — 167

28. About the author page — 168

Suzy Lowry Geno

More Simple Times at Old Field Farm

Shadow

I have fallen in love.

He's attentive. Patient. Protective.

He understands my need for all these farm animals and he's right there ready to help with each step I take as I do my chores.

He eats EVERYTHING I prepare for him and never complains about the menus! He doesn't mind that I'm a messy person and he would rather me be wearing jeans and boots than skirts and high heels! He's still a little bit smelly but I can overlook that for the time being because he overlooks all my faults!

I wasn't prepared to fall in love with such a male of his type because, after all, from the time I was little was always known as a "cat person." But now this huge male dog that is bigger than some of my pygmy goats has waltzed into my life and captured my heart!

He looked like a tall skeleton as he roamed the highway in front of my farm. He basically kept within a two mile area between a small cemetery and the old mining community of Taits Gap. Rescue groups tried to catch him, and one friend was almost run over as she tried to minister to him, but he didn't trust anybody. No human trap could hold him. He was just too big.

Folks left feed at various spots on the side of the road trying to keep him from starving. I put him on the local radio Swap Shop and on Facebook as we searched for his previous owners. Several folks called about him and a couple even came to see if this was their missing dog, but none claimed him.

The weeks dragged on and he looked even more pitiful. He would stand in the highway at the end of the driveway to my little general store and listen to the roosters crowing with a faint smile on his face. My heart was breaking for him, but I was afraid to feed him here on the farm because I didn't know how he would react to my free-range chickens and ducks and all my goats and other animals.

I had a little hen that hatched three small chicks in a plastic box in the corner of my house's front porch. I woke one morning and the big dog was lying on the cool concrete with the little hen and her chicks happily beside him! He ate a little feed I provided him that day, but he wouldn't let me come near him.

The second morning he was again sleeping beside the little hen and her family. I called him and he came around to my carport and ate again and drank the clean fresh water I left out. From then on my carport was his home. But days went by and he still wouldn't let me touch him.

Finally one day he simply walked up to me and placed his giant head against my hand for me to pet him!

Many people thought he had the mange because most of his hair was falling out and his skin looked awful. But I carried photos to Dr. Joshua Standridge and he believed the main problems were from fleas and from malnutrition. A local dog rescuer provided his first flea pill; someone else helped buy the high-quality, high-fat, high-protein dog feed he needed and Dr. Standridge prescribed an antibiotic and a steroid.

I also fed him a dozen scrambled eggs topped with Amish-cheese every day!

The first two weekends he was here, there were terrible thunderstorms and he was not happy as he cowered on the carport. So I sat with him til the wee hours singing lullabies that my kids were rocked to for many years....He would lie at my feet and "smile" at me thumping his tail on the cool concrete floor. He'd get this puzzled look on his face as if to say, "Bless her heart. She can't sing. But I know she's trying to help me!"

I called him "Big Puppy" because he would make little whiny noises like a puppy when he wanted attention. He didn't do much those first few days but eat and sleep sleep sleep.

He'd been here about a month when he came around the house and walked with me through all my morning chores. He matched every step I took so I told him he was like my shadow. So his name officially became Shadow Big Puppy.

Shadow looks to be a Great Pyrenees, a breed of dog I had planned to get when we ever finish fencing our bigger pasture here on the farm. According to Wikipedia and the Internet site for the American Kennel Club, what we call a Great Pyrenees in North America should not be confused with the Pyrenean Mastiff.

The Great Pyrenees developed in the region around the Pyrenees Mountains of southern France and northern Spain with the first descriptions of the dogs dating back to 1407. While males grow to 110-120 pounds and 27-32 inches, and females to 80-90 pounds and 25-29 inches, according to the Great Pyrenees Club of America, the Great Pyrenees is naturally nocturnal and aggressive with any predators that may harm its flock. "However the breed can typically be trusted with small, young and helpless animals of any kind due to its natural guardian instinct."

More Simple Times at Old Field Farm

Shadow now walks regularly through my free-range chickens, ducks, and turkeys, and is comfortable around other animals who visit from all areas here on the farm. He has rubbed noses with the pygmy goats through their fences.

There's also a special rooster, Doc, who belongs to my son, who regularly visits with Shadow on my carport and who eats with him out of his bowl!

Shadow especially loves small children IF they approach him quietly.

Shadow has gained weight and his hair has grown back. He still has some rough spots and he refuses to let me trim any of his hair.

A Great Pyrenees is said to live normally for 10-11 years. I have no way of knowing how old or young Shadow is.

A neighbor remarked that it would be "so sad" if I cared for Shadow and then he died soon of old age. That WOULD make me sad. BUT I cherish each moment and each day I have with this big furry fellow.
   I have written this before, but I believe it even more now. "We are told that we need to be careful in this life because we may be 'entertaining angels unaware.' Sometimes in my fleeting thoughts I wonder.... just for a minute...if some of those 'angels' might just be wearing fur..."

Professional Porch Sittin"

It has been said that all the disagreements in the world could probably be solved if the two main antagonists were forced to sit on a Southern front porch on a lazy afternoon with a soft breeze blowing and with an unlimited quantity of iced tea!

I'm a porch sitter from way back (more on that later) but I hadn't really thought about how that simple idea was storming the country as folks struggle to make sense in this tumultuous world.

It seems Claude Stephens, director of an arboretum in Louisville, Kentucky by day and a porch "sitter" every afternoon decided in 1999 to found the Professional Porch Sitter's Union Local 1339. His group had no motto, Just a "suggestion:" "Sit down a spell. That can wait."

There were no rules and no dues and if you wanted (or still want) to start your own group, all you had (have) to do is pick a number for your "Local" and start sitting!

After Stephens' idea was featured on National Public Radio's "All Things Considered," Professional Porch Sitting groups began springing up across the country. For instance, there's Local 213 at the Mountain Thyme Bed and Breakfast in Arkansas that specializes in "piddlin' around." They state simply, "The best way to do nothing well is to make sure it is done slowly."

Local 150 in Florida even has a Facebook page! But don't think Professional Porch Sitters are only in the south! There's a really active group in New Jersey, Local 1893, that even hosts "membership drives" to get others sitting on their porches, front steps, stoops or whatever they have! Their motto is simply, "Rise up and sit down."

History shows us that most everybody had front porches until about the 1950s when modern air conditioning was being so improved that even the common man could afford it. Until that time, most everybody in the South moved to the front porch in the afternoons to try and escape the oppressive heat.

How many people remember sitting on the front porch shelling peas in the afternoon or breaking green beans?

I have two really early memories of front porches in my childhood. The first involves the front porch of the white farm house just down the hill from my current farm and now owned by the Tidwells.

Shortly before I was born, the road in front of the house (now State Highway 132) was changed to run BEHIND the house. So the interior of the house was rearranged and a new front porch was built on the "new" front of the house and the old front porch, which was now on the rear of the house, was closed in to make a small bedroom for me. That new front porch had a concrete floor painted red and always felt cool to my bare feet. I know there was a metal glider, but I'm not sure what other porch furniture was situated there.

Every Sunday morning (I guess so I would stay out of my mama's hair while she attempted to get ready for church and have a noontime meal ready for our return) my daddy would sit on the front porch with me and read a Bible story to me from his big, black-leather-covered Bible. Nine times out of 10, I asked for only one Bible story: Jonah and the Whale (now I know it was likely some other sort of big fish!). I'm sure he could almost recite the verses out loud since he read it to me just about every Sunday...that porch was hallowed ground to me!

They eventually screened in that porch (while I was still a tyke). Another neighbor would come and wash at the creek adjourning our property (which is now just down the hill from my farm and STILL adjoins my property---just from the opposite direction).

While Flossy boiled clothes in an iron wash pot and then rinsed them in the creek, Conette and I would play in the yard. Our front yard sloped toward the porch from the highway. The front screen door featured the design so popular in the 50s, a metal swan and other squiggly metal designs to protect the screening.

As we raced down the front yard, I fell into that swan, eliciting the first stitches I ever needed on my forehead! (But I still have fond memories of that porch!!!)

(My little general store's front porch is great for porch sittin'.)

More Simple Times at Old Field Farm

The second porch of my early memory was the front porch of my Granddaddy and Granny, Jim and Vennie Inmon, who by then lived on a homestead lot IN TOWN. The front porch featured a blue-painted wooden floor with a white railing all around where Granny kept an unbelievable amount of flowers growing in clay pots AND any other kind of can or jar she could collect (after all she did open and run the first successful flower shop in Blount County that still exists under another name and different owners today)!

There were at least two meal rockers on that porch and a green metal glider. That was where the family sat during the mornings and during the late afternoons to escape the heat. We played countless games looking at cars moving up and down what is now Alabama 75 (such as who first spots the most blue cars, etc.)

After lunch, I'd be drowsy listening to the clank-clank-clank of Arthur Tidwell's old basket-making factory just across the highway and down two lots. And when I was old enough to safely cross that highway (no one would let a child do that there today!) to get my grandparents each a big Pepsi from the tiny gas station across the street, I thought I was grown!

I've tried to have at least some sort of front porch on every house I've owned through the years.

And many of you know, when I dreamed about moving my farm's tiny general store to a building beside my house several years ago, in my dream it had a tin-roofed porch! We built that porch when I moved into the tiny store later that year. The following year, I screened it myself using my miter saw and following the plan my late husband had used screening in my mama's carport!

You wouldn't believe all the porch "sittin'" that goes on there now!

And just as I'm completing this article in my little office, my grown daughter Jannea came in to tell me that she and my huge rescue dog, Shadow Big Puppy, had quite a treat this morning! As she was sitting on my home's front porch, a fawn still with its spots came slowly walking up the fence line across the highway as casually as if it owned the neighborhood! You don't see that sort of thing sitting indoors glued to the TV or playing video games.

When writer Michele Norris in a July 28, 2006, professional porch sitting article declared, "Sitting on the porch is not a place, but a state of mind," she pretty much summed up my feelings.

(Old Field Farm General Store and it's "sitting" porch!)

So even though it's Thanksgiving month while I'm writing this and the weather is getting a little cooler, put on a sweater, pull up a chair on your porch (or deck or anywhere nearby) and just SIT!

You may be surprised at just how good it makes you feel!

Suzy Lowry Geno

The Roosters---Ninety and Nine...

When we first ordered our original 25 Golden Comet hens from the Blount County Farmer's Exchange (Cooperative), Roy and I were often spending hours just sitting out in our lawn chairs watching "chicken TV."

Those first Y2K chicks almost immediately increased by five when I ordered five Americauna (Easter Egg) chicks as well. The bright-eyed little biddies were certainly different than the thousands of squawking broilers Roy's mama raised at her commercial poultry houses years ago as our little free-rangers scratched in the grassy areas gulping down bugs and worms.

I tell everybody that chickens ARE addictive!!! We soon found that chickens were like the old adage of eating potato chips: while you can't eat just one chip you can't stop wanting more and different varieties of colorful little birds! The next year of course I ordered more chicks. But it wasn't long before one began to act a little differently and I couldn't quite figure out what was wrong. It was a little skinnier and it's tail feathers came in long and full.

Then one morning I heard a wavering eerie sound---but it was definitely a first cock-a-doodle-do! That wasn't a weird-looking little hen---we had our first rooster!

Some folks suggested we get rid of him, after all he wasn't needed to make eggs. Others said he'd make a nice addition to the stew pot. But he quickly proved his worth. He sounded a unique call if a hawk flew over and the hens immediately raced to the safety of the coop or the shade of the oak tree, whose branches we left low for good hiding spots. He's also let forth another special call if he scratched up a wiggly worm or another tasty morsel.

Just about every year after that at least one of the new "hens" we ordered turned out to be a little male. And then one of the Easter Eggers hatched about 15 eggs underneath the back porch including three of the prettiest brown, gold and white roosters you've ever seen---with one featuring the grayish, green leggs of his Easter-egg mama!

Most of my "chicken learning" came from what I'd read in books, magazines (such as Backyard Poultry and Countryside), and from the Internet. It seemed roosters fount each other and were really territorial. Lots of folks on the Internet homesteading forums even talked about their roosters spurring them and causing other problems. But evidently my roosters didn't peruse the internet nor read those same books! They pretty much all got along. Sometimes there was a little chest bumping and a lot of cackling if one got to loving with one of the others' girls but no serious confrontations ever occurred. And none ever attacked me.

But Roy said what happened every night just as it was getting dark was what was truly amazing to him even though he'd been raised on farms throughout his childhood.

My chicken coop has been added onto four or five times, so there is one door at the front and two doors on each side.

Each night each group of hens lines up with "their" rooster and marches into their individual doors to their partitioned sections of the coop. Roy said it reminded him of boys and girls lining up behind the flags and marching in to Vacation Bible School at church when we were all little.

But the most amazing part happens on some nights after that: maybe a chubby hen slips into our dog Maggie's fenced area because there is some choice worms under her pens' thatchy grass and she lets time slip away. Or maybe another hen accidentally is pushed into the wrong door by a group of girls rushing in to get the choicest spots on the roosts. Or maybe another just tarries too long out by the feeding trays.

I don't think roosters know how to count but evidently each rooster knows exactly how many hens are "his" and each knows when every one of their little harem of 20 or more is missing!

That particular rooster comes trotting back out the door, leaving the other hens inside, but clucking, calling, and looking in the bunny barn, under the cedar tree, in the edge of the woods, until he finds his missing hen!

He then escorts her back to their door, scolding her all the time, until she is tucked safely inside with his other girls.

Whichever rooster it was (or still is) that finds one missing won't rest until each and every one of his charges is safe inside for the night. And this farmer better not try to shut that door until EVERY ONE is accounted for!

I've often said that some of the times I've felt closest to God have been when I've been around my animals: witnessing the miracle of birth or even just the calmness of a barn smelling sweetly of hay as rabbits and goats munch their evening meals.

So the roosters have set me to thinking of how much more our Savior seeks us out when we stray from His fold of safety.

Jesus told a parable of a shepherd who had 99 sheep safely in the fold (Luke 15:4-7) but who left them to seek out that one tiny lamb who had gotten left behind or who had strayed too far away from his safe haven.

Children's Bible storybooks often feature an illustration of a kindly shepherd striding back toward the flock carrying that single lamb draped across His shoulders.

It's a SIMPLE story with a SIMPLE premise: the rooster going back out after one errant hen or the shepherd seeking one lost lamb.

It won't qualify as the theme for a mega move nowadays or the plot of a bestseller.

Suzy Lowry Geno

And you may think it strange and not understand why this simple woman is musing about such things on her homestead: unless you've been that little lost lamb...

Tired but not "retired"

I really didn't know how to respond when a lady last week asked me how I was liking being "retired."

When she asked me I had mud and chicken poop well toward my knees on both legs of my blue jeans, my shirt was plastered with sweat and my hair---oh that gray hair---was kind of straggling down my back held somewhat in place by a big, plastic hair clip that belongs to my youngest daughter.

I'd just stuck a nail in my wrist as I was struggling to get some errant hens out of the guinea pen. I'd stepped in an armadillo hole while surveying the back goat fence for storm damage AND I'd managed to mash my finger while using my drill to screw a piece of plywood back onto the front goat enclosure.

I THINK I was polite when I told her that, while I don't know have a full-time newspaper reporting job, I am definitely NOT retired!

When I posted something on my Facebook page about the retirement question, I got all sorts of replies from folks who are always interested in my somewhat unique lifestyle. And it doesn't matter who or what interesting thing, farm, business that I write about, the SIMPLE TIMES column that provides the most feedback through email, phone calls, texting, and other means of communication are the ones I receive after I write about my simple little homestead or what I've been doing here.

One of my 12-year-school classmates asked where I got my energy to get everything done. I replied somewhat comically that "poverty is a great motivator." But that's not completely it.

I was reading a "Report from Them That's Doin' " from the November/December 1977 Mother Earth News last night and the couple quoted a question from a book called "Working Loose," published by Random House in 1972. They said, near the beginning of that book, it asked, "Forgetting about money, what would you like to do more than anything else?"

As they pondered that question and reevaluated their lives, they left the corporate world and raised their children on a farm whose money crop was primarily honey! While their so-called standard of living was "lowered" and their annual income was as well, the man no longer had to take antidepressants AND a strange rash which had been plaguing his body off and on for months suddenly went away!

Just this morning on the Internet, I was sent a photo whose caption read, "Don't be so busy making a life that you forget about LIVING!"

Yes...I know. In this modern world, we can't completely "forget" about money! (Believe me, money has really been on my mind this morning since I have to buy a new battery for my truck!)

But as I said in a previous article, which received so much feedback, if we are simply working to make money to pay for all the latest things for our homes, cars, etc. yet we are so busy working we are seldom home to ENJOY many of those things, then something is incredibly WRONG!

Somebody told me that was too much of my old "hippie" philosophy coming through. No, I prefer to think of it as a part of my Christian walk...

There's a small group of young men who are making videos to show on the Internet for free and one of the most recent showed random people who were asked to share food with those who said they were hungry...in nearly every instance it was the homeless or the "less wealthy" who immediately shared what little they had, often dividing in half the one sandwich they had just been given by someone else!

All of these things make me stop and think. There's another old saying (and I'm not sure who deserves credit for this one) that says, "Find something you love doing as your career and you'll never WORK a day in your life."

I guess that's where I'm at in my life right now. I "work" 10 to 12 and even 14 hours every day, but I wouldn't do all this if I didn't truly love it.

A typical day for me begins by me feeding and watering the eight goats, 400 plus chickens, 30 plus ducks, 10 rabbits, two dogs, five cats, five turkeys, and about 15 guineas.

Any animal that needs special attention gets it at that time such as medication or wormer for some of the guys and gals and possibly hair trimming for a rabbit or two.

If anything needs to be nailed up, replaced or wired together, I try to see to that then as well.

I usually open my little farm store, hang out the flag, label any jelly and soap that needs to be stocked in the store, order any Amish-made products I'm getting low on, then make any jelly that needs to be made that day. On days I'm not making jelly, I usually make goat milk soap. It is still hot right now and I am thankful that currently there are no goats that need twice daily milking!

I gather eggs from the hens around lunchtime and at least twice more during each day, clean them, and place them in the little store refrigerator. Any time I sit down there's a quilting project or a knitting or crocheting project nearby so even time on the phone is not wasted. (These are usually items for sale, not just for fun!)

If I get up really early (often before daylight), I try to do whatever writing I can before my mind and my body gets cluttered with more physical chores.

During just about every season there is something to cultivate, water, propagate, pick or gather. A recent project has been to make gallons of what we call "Granny Soup" filled with whatever vegetables I have on hand which will feed me well into the cold weather months. There's nearly always something drying in the dehydrators for either dried herbs or herbal teas.

During these fall months I hope to get my raised beds finished in time for this coming spring's growing season.

Each season has its own specific chores. In the winter, I carry in enough wood in the mornings to last throughout the day and, each late afternoon, I carry in enough to last throughout the night. Then there's taking out the ashes and extra sweeping required.

Hopefully there will be more time this winter to work on continuing to learn to spin not only my rabbits' wool but sheep wool and other fibers!

I do just about what all the "big" regular farmers do, except I do it on a smaller scale since this is a "one woman" operation. While I don't bring in large sums of money, my profit margins were figured once by a financial guru and those profits percentage-wise at least were in line with much bigger farmers.

Helen and Scott Nearing, long considered the "parents" of the back-to-the-land movement in the 50, 60s, and 70s, divided their waking time into sections: for hours of "bread" labor (work to provide for themselves on their homestead); four hours of labor to earn monies that were necessary (they were both writers and speakers); and four hours used to help others.

While I don't agree with the Nearings' socialist leanings, I do agree with their divisions of labor. Even when I worked full time for a busy daily newspaper and also for a weekly county newspaper, I never ever worked in an office other than my home office. I guess I've just never been a nine-to-five type of gal.

And as for retirement: I don't plan on having one!

Hopefully I won't have to stop working until I'm laid peacefully in the ground on that hill atop Oak Hill Cemetery beside my husband.

Suzy Lowry Geno

More Simple Times at Old Field Farm

Shadow Big Puppy didn't know quite what to think about a big snowfall on the farm! Weather can cause my days to be a lot lengthier as a struggle to tend the animals in the snow, mud or even heat!!!

Smells take me back....

I was just watering the more than 300 heirloom tomato seedlings in my little greenhouse. The sun had beamed down upon the tiny plastic-covered tent-like playhouse-sized building just enough to make it pleasantly warm as a myriad of unidentified insects buzzed about my head.

But as I happily went about my chores I was stopped in my tracks as so often happens around this little homestead. The smell of the water stirring the soil in the little cups, the sweet acrid aroma of the little plants themselves, and even the dampness of the crunch gravel beneath my feet immediately took me back about 58 years...

My Granny Inmon owned the first florist shop in Blount County. By the time I came along she had a neat little building beside her then-house along Highway 75 just south of downtown Oneonta. At the rear of that little building was a "hot house" room built almost as an afterthought. And that little room was where I was transported in an instant with just a strong sniff

inside my little greenhouse today.

It happens to me all the time. I can be using the miter saw to cut a post or a piece of wooden trim and one whiff of that sawdust and I'm a little girl once again...Running into my daddy's arms as he arrived home from work, wearing his overalls with his nail apron stretched across the front. His smell wasn't of any fancy aftershave that was advertised on the radio. No---his fragrance was one of hard work: sweat, sawdust, and the leathery smell of the back brace he had to wear before miracle surgery made him better!

And then there's the 36 baby chicks currently residing in a big cardboard box under a heat lamp on my back porch. Is there any sweeter smell than that of warmed baby chicks scrambling about on wood chips as they scratch their chick starter feed and splash their water? We lived "just down the hill" when I was about four and my brother Bobby, who was nine years older than me, raised a flock of fluffy yellow chicks in the old chicken coop as one of his first FFA projects. He held me up to the brooder, my cheek resting on his soft, often-washed FFA blue denim jacket, and the warm smells lingered in my memory as I petted a soft ball of fluff.

And just about nothing can compare with the smell inside my little bunny barn. Right now there are ten bunnies (five for fiber for my spinning wheel---they are sheared not skinned!!!)

and five sweet to me! When I walk in and smell their little furry warm bodies munching on fresh hay even the rabbit pill manure smells sweet to me!

I can walk into the back of the Blount County Farmer's Exchange or any of the co-ops across the state, and the feed smells take me back immediately to Clarence Sellers' small store where brightly colored fabric bags of flour nestled on one side of the storeroom while fragrant feed for just about every animal imaginable lined the other shelves. Or the smell of seeds in the big old time bins at Lowry's Store in downtown Oneonta, owned by one of my Grandpa Lowry's brothers!

I worry about kids today. An entire LIBRARY of books can be included in one hand-held electronic tablet or Kindle reader, which seems like a miracle to me. But they are missing out on one of the most wonderful fragrances in the world: that of a new book.

Can you remember those smells on the first day of school? Those new textbooks, that freshly-sharpened yellow pencil (with maybe an extra red rubber eraser with its own unique smell!), all mingling with chalk dust from the blackboard and that green stuff sprinkled upon the wooden floors by the school janitor in order to sweep the dirt carried in by hundreds of little feet back outside?

And what about the smell of those little new white Bibles all of us little girls had in the 1950s bought for us by our parents from the downtown "dime" store and likely given to us as part of an early-morning Easter basket? Some had zippers all around to help close them, featuring a tiny golden cross at the end. Someone recently gave me a 1950s era white Bible they'd found while cleaning an aunt's house after she passed away and the very first thing I did was open it about half way and sink my nose into the pages somewhere between the colored pictures of David slaying Goliath and the manger scene with baby Jesus! Yep! It smelled just like my Bible did when I was a little girl!

Before you think I've dropped completely off the dark edge this time, here's what the experts are saying: "In an article called "How Smell Works" by Sarah Dowdey on the website http://health.howstuffworks.com she explains, "Because the olfactory bulb is part of the brain's limbic system, an area so closely associated with memory and feeling, it's sometimes called the "emotional brain," smell can call up memories and powerful responses almost instantaneously."

She says we are "tightly wired" in that the "olfactory bulb has intimate access to the amygdala, which processes emotion, and the hippocampus, which is responsible for associative learning. But she also notes that smell would "not trigger memories if it weren't for conditioned responses."

But it can work both ways: for instance, a carnation can make you think about your boyfriend buying you a carnation corsage for your spring dance, conjuring up happy memories. OR the smell of that same carnation can remind you of the funeral flowers of a dear friend, which makes you sad.

Happily most of my "smell remembrances" are of good things. Dowdey says since so many new odors are encountered while we are youngsters, many smells often bring us to childhood memories.

Who doesn't love the smell of soil that has freshly been turned, whether it's an entire field or garden or simply a window box.

I can remember riding on a "slide" being pulled behind a mule as my daddy flattened out clumps of dirt in our hillside garden. I was barely three at that time. There's the smell of freshly ironed clothes as my mama sprinkled more water on the next shirt as she pressed it into submission.

How about coal in those old round coal burning heaters?

Evening in Paris perfume sitting on the old dime store counters for sampling?

Bolts of fabric just lining the counters on the second floor of the only store in Oneonta that HAD a second floor!

Those memory smells come from a time when life WAS much more simple...and that's probably why I (and many other back-to-the-landers) strive to get back to that simple time...

Many folks think I'm nuts. Others wonder why I work so hard. I know there's lots of times even those in my own family have questioned my sanity.

But as my grown son Nathan, who lives nearby, put my animals "to bed" for me last week he learned something from the smells, sights, and sounds of the late night homestead that has helped him to understand.

Here's what Nathan posed on Facebook later that night: "As I put mom's animals to bed last night (don't worry, she's fine, just at church), I may have caught a piece of understanding.

"With my usual sensory overload style, with this moon overhead; I could hear the chorus of frogs across the pasture- when I say chorus, I mean almost siren-like fall of voices.

"I could smell the warm earth, and grass, cooling just after the sun's setting.

"Hens inside the hen houses softly cooed and clucked as they settled in for a cozy sleep. The guineas chur-chur-chur as they try to select their perfect limb to roost on in the nearby pear tree.

"Goats gently bleat at me as I close the doors and gates against whatever the night may hold.

"I can hear a rabbit in the "bunny barn" lapping at its water bottle clackity clackity clack...

More Simple Times at Old Field Farm

"I'm usually well on my way to bed at this time of night, but tonight I'm grateful. I've seized a little bit of the peace mom gets to have every day.

"Farming is never easy, or sometimes profitable.

"Every day however, There IS a harvest to be reaped..."

(I was blessed that the previous article, "Smells take me back" was chosen not only the Class #12 First Place Feature Article in the annual contest sponsored by the National Council of Farmer Cooperatives during their Information Fair, it also was awarded BEST IN SHOW of all articles for 2015!!! Here is a photo of my son, Nathan Geno who contributed a great deal to this article!)

The next column, "Happy You Happy," was awarded Honorable Mention in Category #12 for 2015 as well. I was so blessed to have double honors this past year!!!

Happy, YOU Happy...

"Happy, you happy," she said as she reached out and touched my arm.

I can only understand some words in Spanish and she is struggling to learn English but we did not need her teenage daughter to translate for us as we shared that goodbye phrase and smiles.

The beautiful young mother with three children traveled here to Old Field Farm from another county to buy "farm fresh eggs from happy chickens" from my tiny general store.

Her four-year-old son's bright smile would have powered a solar system as he wandered amongst the free-range chickens, ducks, guineas, and turkeys. And the funny "nom nom noms" coming from the goats as he and his little sister fed them peanuts was likely the highlight of the afternoon.

They all peppered me with questions. When the mother learned I managed the farm and do all the farm work myself, she fired off questions to me almost faster than her teenager could relay them to me! And she understood most of my answers, laughing and enjoying one of the sunshiny days of winter.

It was one of those visits that make me appreciate my little homestead and thank God for my many blessings here. And it also set me to thinking...

I am likely the happiest I have ever been in my life. Not a ha-ha-ha type of happiness. But a feeling of true contentment. That may seem odd to some people.

In the last few years I tended my mama and my husband and then sat at each of their bedsides as they passed from this life to the next. I lost my mother-in-law in between the time of my mama's and my husband's deaths. And I can guarantee to you that those kind of losses will change your perspective on life forever.

My finances seem to be an ever-present uphill battle (which I am conquering one paid off bill at a time!)

But oh my goodness! I am a woman who made her living most of her life through words. But there aren't adequate words to explain what I mean by all of this. But I'll try.

One of the simple living magazines that make their way to the used rack in my little store, had a whole section in January 2011 about how to make that year one's happiest ever.

That author began by talking about how Thomas Jefferson's penning of the Declaration of Independence began by noting that "all men" have a right to "the pursuit of happiness." And they further explained how many folks erroneously equate that kind of happiness with "success."

But then several different authors on multiple pages noted one could achieve more happiness by "chilling out," taking naps every day, remember "whence you came," and more, including make your bed every day! (oops I'd likely really fail that last one!)

Then one began talking about looking out over a European city at a huge cathedral and how that made her feel even though she admitted she was not Catholic and "not even a Christian."

Well, everything in that article just went downhill from there!

There's an old saying that goes something like "you can't have true happiness and contentment if you don't know where you've been!" Well I can paraphrase that saying to note that "you can't have true happiness and contentment if you "don't know where you're going!"

Folks that know me know I've made some humongous mistakes in my life, many many many times And if my future depended solely on my own thoughts, my own actions, my own feelings, then I would be in pretty bad shape. But I'm in pretty good company!

Paul (then known as Saul) not only just persecuted Christians in New Testament times, evidently he was in charge of some of them being put to death by stoning and worse. But then he had an encounter on a rural road and his life changed forever.

And I love what he talks about in Philippians 4:10-13. He talks about being content in whatever situation he found himself. The Greek word for "content" can be translated as either "to be self-sufficient" or to be "satisfied."

So I guess that's what I mean when I say I'm happier than I've ever been in my life. I am "satisfied." This morning on the computer I was texting back and forth with one of the young mothers from my church. She, her husband, and her four children are getting baby chicks for the first time and I have been helping by explaining what they will need to care for them, what to expect, and "heavy" topics such as the pros and cons of having a rooster!

It makes me feel good to not give advice but to SHARE what I have learned on my homestead journey during these years past. Often what I am sharing is from making mistakes and then knowing what NOT to do the next time!

That is part of my happiness, my contentment.

But my true contentment really has nothing to do with this little homestead; nothing to do with these animals I cherish so much; nothing to do with how many jars of jelly I sell or how many hours I rock on my tiny store's porch.

That happy message was simple more than 2000 years ago and it's simplicity now may be what confuses folks so much.

True simple happiness, true simple contentment, comes only by Grace through Christ Alone...

So yeah, you can truly say that I am "happy, happy, happy."

Weird again...

It happened again this week. Someone found out I do not now own a TV set and they almost insisted they were going to bring me one!

The couple looked at me with almost an expression of horror on their faces as I explained I don't have one because I simply don't WANT one.

Sure economics entered the picture when I first decided to cut the ties with my cable TV company. It seemed every month my bill was going higher and higher, but I was finding less and less on the channels I actually wanted to view.

Most everybody in my age group can remember when TV was something that came free over the airwaves. You bought a TV, hoisted an antenna and watched maybe two or three channels that were available.

Even that changed a few years ago when new regulations went in affect and now you need a TV equipped to work with one of those little antennas you stick in your window with a suction cup. You can get several more digital channels it seems with those things. But somehow my homestead's location prohibits even those from working unless you install it above rooftop level.

When my husband died, the HUGE TV he insisted on watching was about to go out. SO it's now with my son. The other two older model TVs in the home were so big, so old and so heavy I couldn't even move them around, although they both still worked well. I sold one for $40 and gave the other one away!

But I'm not one of those folks who have sworn off TV completely for religious or other reasons. Right now anything I watch is either on my desktop computer in my office or (what I use most often) a small eight-inch tablet that keeps me connected to the world!

But I'm not alone! There's even a term for those like me. The great survey firm Nielsen call us "Zero Households." They note there are now 5 million households (of the current approximate 125 million households in the United States) who currently don't have "at least one" traditional television set in their homes.

So I'm one of the less than 5 percent of U.S. Households without a traditional TV, but they say that number has doubled since 2007. But Nielsen explains on the Internet, "Being a Zero TV household doesn't mean there is no TV. They are connected to the Internet, not to cable or satellite services."

On my little tablet, I can keep up with the local news and weather via the Birmingham-located TV stations thanks to free apps I just downloaded from the Internet. I watched every Alabama football game but one, thanks to special sports apps and TV network apps.

I have watched every episode of "Simply Southern" (the wonderful TV show created jointly by Alabama Farmers Cooperative and the Alabama Farmers Federation) by going to their YouTube channel.

YouTube has become a wealth of entertainment and knowledge as I regularly watched shows produced by other homesteaders such as "An American Homestead" and "Appalachian's Homestead with Patara." And there are thousands of videos free for the watching on how to make soap, how to quilt, how to can, how to do counted cross stitch, how to tend to all types of animals and so much more. And they're not dull! Although many are educational, they are usually quite entertaining as well!

Remember Carla Emery, the wonderful back-to-the-lander who in the 1970s wrote the giant book "The Encyclopedia of Country Living"? Her daughter has a YouTube channel (just look up "Fouch-o-matic" on the YouTube search engine) detailing how they're currently living in a yurt while building their dream cabin in the middle of the woods! Their Christmas show was especially precious!

(Oh, and "An American Homestead" and Carla's daughters' families are completely off grid, using solar power to access the Internet and edit the videos they produce!)

And when I find I just HAVE to watch something from "regular" TV (I do like the shows "Alaska the Last Frontier," "Life Below Zero," and "Mountain Men" I can watch them via my son's connection, but that's not too often.

One thing I have discovered is that I can no longer turn on the TV and let its mindless drivel drone on in the background simple because it was "there." I'm not bombarded with political ads, commercials with loud aggravating music and no-substance shows that come on one right after the other.

Streaming shows over the Internet through NetFlex or Amazon or any number of other such companies also lets me pick and choose what I want to watch and when I want to watch it for nominal cost. But it would be easy to get "addicted" to those services as well.

I've seen on Facebook where folks talk about a marathon of watching an ENTIRE SEASON of a certain TV series over the course of a long night or a weekend! I can't imagine sitting through all that!

In his May 15, 2012 article in the Business Inside telling why he no longer had a TV, writer Alden Wicker explained," It's costing you money....You're keeping up with fictional Joneses....It makes you covet things you never knew you needed...Life is more fun without TV....It's cramping your social life..." and "It can make you fat."

According to LifeScience, "about 1 to 2 percent of those in the U,S, refrain from TV all together...Some give it up to avoid exposing their families to excessive sex, violence and consumerism."

But from what I've read and seen, the majority of folks without "traditional" TV service in their homes are more like me---watching things via the Internet (or even just through DVDS).

Right now I'm researching what sort of greenhouse I want to build back since my tiny 5-year-old greenhouse was recently destroyed in a wind storm! There are so many ideas and free videos on the Internet about greenhouses; I will just have to force myself to stop and actually start building!

So for those of you who are worried about me while I live without a clothes dryer, heat completely with wood and have NO TV set, don't fret!!! I'm living the blessed simple life of my dreams!

Poor but rich is what some folks can't understand

Are you ready to pull your hair out yet?

Since I retired from my more than 30 years as an investigative reporter a few years ago, I've striven to stay as far away from politics as possible.

I still pray for our leaders and I certainly try to stay as informed as I can because I believe it is our God-given responsibility to make wise choices from the local all the way to the national levels. As I sit here on my little homestead, it is sometimes almost too easy to isolate myself and my thoughts from what's going on in the outside world.

But, sometimes, there are statements made by some of the candidates that make you want to scratch your head, squinch your eyes and wonder if you just heard what you thought you heard.

Once such statement lately, at least I feel, qualifies to be in that head-scratching, eye-rolling, what-in-the-world-was-he-thinking category.

The candidate made the statement that he didn't feel anyone had experienced REAL poverty except a certain group of people.

What?

In my nearly 64 years of expertise, I've seen poverty across all age groups, all races, all sexes, all religious denominations---in my most humble evaluation, poverty strikes some who have almost literally "worked their tails off" to others who "wouldn't hit a lick at a snake," a couple of phrases from my rural South.

I think what that particular candidate, and many others who are isolated above and away from normal folk, misses is that even if poverty, hopefully at least temporarily, is someone's lifestyle, it is their state of mind---and the state of their hearts---while in that situation that makes all the difference!

My great-great-grandmother traveled to Alabama via covered wagon when her well-to-do- Illinois family didn't like her young boyfriend. So they eloped and set out with some of his family for a better life in the foothills of the Appalachians.

They'd travel for a while, then stop in a community to work for a season or two, and began to have little ones as the years progressed.

Along the way that great-great-great granny also began experiencing something that has plagued many of the females in my mama's side of the family for decades: she began to lose her sight!

But did she sit in the wagon, hungry, cold, often-pregnant and cry "woe is me!" Nope! She cooked their meals over an open fire and was the soul who kept her little family together!

From my mama's written records, "After they got to Alabama, she was totally blind, but she did her own cooking, sewing and everything else...She would keep the kids for the others to work in the fields and have their dinner ready for them when they came home.

"One day when they were all in the fields, she had to go to the outhouse. They had been digging a new well and she fell in. It didn't hurt her, but she had to stay in that well until they came in from the fields and heard her!"

One of her daughters, Sofronia (Fronia) Dover married Mark Windsor.

Sofronia (Fronia) Dover Windsor

Here's another bit from my mama's writings about her Granny Fronia. "Grandpa must have been in ill health all the time because Granny worked all the time to make a living. She took in washing for people. She'd stand out in all kinds of weather at a spring and wash for people (using a big black, iron wash pot to boil the clothes she stirred with an oak paddle to make the lye soap dissolve.)

"Sometimes she would heat a rock to stand on to keep her feet warm."

"Mama and Uncle Will (two of Fronia's kids) worked in the cotton mills in Alabama City (Gadsden) when they were just little kids. There were no child labor laws then. I remember mama {this author's Granny Inmon) telling about getting her long hair tangled in one of the looms and it was about to throw her into the machinery when somebody fell against the belt and knocked the belt off, saving her."

My Granny, that girl with her hair caught in the loom, got married at the ripe old age of 14. Jim and Vennie Inmon went on to have seven kids of their own. At first they lived in a little cabin that was half logs and the other half a canvas tent! Then they lived in a tent; they moved regularly as Grandpa worked with a crew building roads in and around what is now Snead.

Later he worked for a dairy and then sharecropped with the family moving almost every year. But those seven kids didn't have bad memories, although they were certainly living in poverty! Grandpa always hung a tire swing and Granny always planted flowers around a neatly swept sidewalk area. There were pet cats and kittens and once even a pet goat trained to pull the kids in a cart throughout the neighborhood!

More Simple Times at Old Field Farm

The Depression of the 1930s hit. My mama always liked to watch the Waltons on TV, BUT she noted the Walton's were rich because they had such a nice house, running water, electricity and even a phone! But through it all, my mama's family KEPT WORKING and kept trying to better their lives!

My Granny eventually sold enough flowers that she was able to open a florist shop and they paid cash to build a wooden house right on Alabama 75 near downtown Oneonta. And everyone of their kids, grandkids, great-grandkids and on down the line to hundreds of descendants, now, kept working as well!

Jim and Vennie (Windsor) Inmon

Some are well-off by today's standards; others have had times when they just got by financially. But while some of us may have lived in what some would consider poverty, there has been a difference.

One of my pastors now has a ministry that regularly goes to Nicaragua where TRUE poverty can be seen almost everywhere. And I lived for a time in South Dakota several years ago, near the Pine Ridge Indian Reservation, where again, TRUE poverty is on every corner. And there are instances of older folks, disabled folks and more who are living in poverty because of no fault of their own.

But I feel---and mind you---these are just the reflections of a simple homesteader in the foothills of the Appalachians---when folks I know about experienced poverty here in this wonderful country, there has never been desolation and long-term despair.

There are no governments or other charitable programs that will result in long-term solutions if the main foundations of a contented life are not there: a faith in Christ alone and that all-American spirit of working hard each day.

That may sound overly simple, but, like I said, that's just the opinion of this simple gray-haired homesteader---who continues to work hard each day.

The turkey experiment

My first turkey came to live at Old Field Farm quite by accident.

Tom, now known as Thomas, came to live here with the addition of 10 chickens, one guinea, two rabbits and two roosters when their owners could not keep them any longer.

He had been living with the 10 chickens so I just kept him in an area with them away from my other chickens and Thomas, as well as his family of hens, thrived.

Then my grown son Nathan ordered six day-old Broad Breasted Bronze poults and raised them on his adjacent farm under a heat lamp. When the poults grew large enough and were fully feathered, they moved the poults into an outside pen. As soon as they grew a little bigger, Thomas moved himself from my farm to their turkey pen!

Since most turkey orders have to be "straight run," no one was certain the sex of the other turkeys. But it was soon determined that between us we now had Thomas, four other males and only one female (whom I named Ruby).

Nathan was raising his turkeys for Thanksgiving and Christmas dinners. (Anyone who knows me well knows that none of the animals on my farm will be eaten; they have to provide income in other ways such as the chickens laying eggs, goats giving milk for my homemade soap, guineas providing snake patrol, etc). But Nathan is a more "practical" farmer than I am and the turkeys were one step up from the few meat chickens he had raised.

But sometimes life gets in the way of our plans!

The turkeys developed a respiratory infection because of cold, wet weather and were on antibiotics shortly before Thanksgiving so could not be butchered. And after the holidays, Nathan was extremely busy in his electrical company. So the turkeys just kept eating and growing...

So this spring, often in the afternoons, shoppers to my tiny general store were greeted by six huge, very healthy, happy turkeys doing their drumming sounds, spreading their tail feathers in displays that would make any peacock jealous, and sidling up to folks just to be petted on their weird-looking, rubbery, blue and red heads!

More Simple Times at Old Field Farm

Hundreds of times in the past few weeks I've had adults tell me they've never ever seen a turkey before except those dressed in plastic and ready for sale in grocery stores. Just last week, a very nice BMW screeched to a halt in the middle of the highway, turned around in a nearby driveway and pulled up at my farm as we were burning off a ditch, which was filled with limbs from a recent wind storm.

Windows opened, three heads poked out and they peppered us with questions about our helps; the six very large, curious, turkeys!

And oh my goodness, the turkeys were the stars during the Highway 132 18-mile-long yard sale held each spring on the last weekend of April! I walked out my side door one morning at

8 to find a couple videoing their 5-year-old son surrounded by the turkeys!

And those photo and video sessions continued all day for two days as people stopped from their search for bargains and just had to have their photo made with a strutting turkey!

In E.B. White's book, "One Man's Meat," a collection of his magazine and newspaper columns from 1937-1942, he talks about his one surviving female turkey (out of an original six) He itemized her total cost to him and, even with 1938 figures, it was a whopping $402.85, including everything from 30 cents for the fertilized egg, expenses for the broody hen who hatched her, growing mash, scratch feed, pens, and more!

(White later used his true farm experiences as background for his books, "Charlotte's Web," Stuart Little," and more!!!)

And I fear the true cost of these birds would be much more by today's standards! If Nathan butchered them now, they would probably have to use them as ground turkey simply because of their age and their size! But it HAS been a fantastic learning experience! (and no one is going to eat ANY of these turkeys!!!)

We do know now that IF turkeys are going to be butchered the deed needs to be done when they are between the ages of 18 and 24 weeks. And there are many more breeds of turkeys to look at if you are planning ahead.

I have a friend out west who usually raised two Midget White turkeys every year, naming one of them Thanksgiving and the other one Christmas. She says the taste is so much better than one bought in a store that her family would likely rebel if she served a store-bought turkey!

There's also Standard Bronze, developed in the United States. In the 1700s, colonists began establishing settlements and the turkeys they brought began crossbreeding with the Eastern wild turkeys. It has changed some through the years as they have been bred for bigger breasts and legs such as the Broad Breasted Bronze that Nathan raised.

The Bourbon Red is also an older American turkey from crosses of Bronze, white Holland and Buff turkeys. According to an article in GRIT magazine, Narragansett turkeys were named for the Narragansett Bay in Rhode Island where more domestic turkeys brought from Europe in the colonial days crossed with wild turkeys making good meat, broodiness that was good for raising poults and with calm dispositions. Then there's the Broad Breasted White that has been developed in the last half century and usually raised in industrial farm situations.

Turkeys are not said to be very smart (but I have yet to see one stand in the rain with its head looking upward for so long that it drowns,) which is the common rumor! Oh, but I have found these turkeys to be extremely SMART about some things!

They know that if someone drives into the driveway of my tiny store there MIGHT be treats involved and, if they cooperate while photos are being made and while they're patted on those funny heads, there's almost a CERTAINTY a treat will follow! These turkeys each also know their own names!

Raising turkeys is similar to raising chicks: They need heat lamps when they are small and basically until they are fully feathered; they should have the finer poult feed when they are young; and they should be kept dry and out of drafts!

Turkeys should basically not be housed with chickens because of the possibility of the spread of diseases between the two species although ours have roamed the fields with our free-range chickens late in the afternoons with no bad side effects. (But I would not house them together with chickens unless it was a situation like with Thomas where he originally thought he was a chicken and had been raised with them by someone else!)

Will there be turkeys raised on this simple farm again?

I would guess that is a 100 percent probability! Ben Franklin didn't try to have them named as our national bird for no reason!

They are fun, interesting, and rewarding birds!

More Simple Times at Old Field Farm

I am an Addict...and it's Great!

I make no bones about it. I am an addict. And you might as well not even try to help me overcome my two biggest addictions.

First of all, all my family knows I am a newspaper junkie! And I don't mean that sterile, glowing print you can read on the Internet. I mean honest-to-goodness, I-can-smell-the-printer's-ink, old-time newspapers on newsprint!

My grown kids have known for years if they are anywhere other than our area, they better bring mama back any small town or big city newspapers from wherever they are traveling or visiting! All that knowledge and all that good reading about others across this land sure helped me during my more than 30 years in the newspaper business!

So it was with great delight that I learned one of my cousins, Robbie McAlpine, who owns the Alpine Advertising Agency, had tracked down the Saguache Crescent, a 135-year-old newspaper that still prints in linotype

He'd seen a feature on CBS news recently, called the owner, and subscribed since the out-of-state rate is only $18 a year. Not only is that small Colorado newspaper printed the old-timey way, but a recent front page boasts the per copy cost is still only 35 cents!

The front page also showed news that most any folks would like to read. While there were announcements for local candidates, there were also articles such as the "Saguache Quilters," "The Rocky Mountain Range Riders" and a "Celebration of Ranching."

The Classifieds were equally both educational and entertaining. One wanting office help listed the job classifications: "good telephone voice, good computer skills, long hours, low pay, absolutely no chance of advancement!" And nearby was an ad for a "free beautiful rooster."

I can't afford to live in Colorado, but that newspaper sure lets me know "my" kind of folks like it there!

And then there's my other addiction and it's equally as bad.

I can't pass up beautiful fabric, particularly old-time cotton fabric such as that from feed sacks, old aprons, and more. My home office, which doubles as my sewing room, usually looks like a fabric store was hit by a tornado and the remains all dumped between these four walls.

I used to love to go to estate sales. I often found boxes filled with unique items such as quilt fabrics which had already been cut to sew, spools of thread often with a threaded needle stuck into its roll, and so much more. I always tried to finish each project, aware of the loved one someone had lost who was no longer able to sew for their family.

Now I am fortunate that several people understand my addiction and they help me by providing just what I need by bringing me such goodies whenever they have lost a loved one, or are simply downsizing. Such was the case last weekend. A very sweet couple brought me several pieces of cotton fabric, but the neatest items were stuffed into a throwaway plastic bag from a grocery store.

There were about 60 paper-pieced quilt squares! And the best part was that the NEWSPAPER was used to sew that those pieces into their designs and was still attached to the back of the squares! I was in addict Heaven.

The paper-pieced quilt squares had 1956 and earlier newspapers as the backing!!!

More Simple Times at Old Field Farm

I started reading all the seven-inch squares and soon found I had clippings from three newspapers from January, February and March 1956, (when I was a dainty 4-year-old)! These newspapers show how our lives have gotten so COMPLICATED since then. Although there have been improvements, somehow I long for that more simpler time!

I know the average income was way below what it is now, but it just seems those dollars went further. Here in the Alabama Farmer's Bulletin from early 1956: "80 acres with a seven room house, electricity, barn, outbuildings, 'lasting water,' several pecan trees, school bus and mail route. $11,000, Lee County."

Or another in DeKalb County: "40 acres, 28 cultivated. Five acre pasture, balance cotton. Six room house, electricity, poultry house, barn, two wells, orchard, $75,000 or exchange for farm with two or more houses."

Or this interesting ad: "Farmall" A tractor with cutting harrow and turn plow, Ford tractor with Bush and Bog, cutting harrow, turn plow, planters, cultivators, mower, $900 or exchange for 3 young Shetland mares in foal."

And these two would certainly be "illegal" now unless you complied with lots of rules and regulations, but then you could get: "vegetables, pickles, 25 cents per quart or 12 for $3" from an enterprising woman in DeKalb County or "All kinds canned green vegetables, 12 quarts $4, 48 quarts, $15. Apple, pear, and blackberry preserves, jams, jellies, 50 cents per pint" from another hard-working homemaker in Jackson County. Oh, and that year the going price for most hay advertised was 50 cents per bale!

Some of the other prices in the weekly newspaper clippings from our area showed men's tee shirts 3 for $1 in Arab and

Chenille bed spreads for $2.98 to fit a double bed. Oh---and something you don't usually see any more" Men's dress hats were $1.98 to $2.98!

Bath cloths were five cents each and ladies' panties 5 for $1!

The grocery store ad noted "young tender pole beans, 2 lbs for 29 cents; yellow squash two pounds for 19 cents, and cabbage four cents a pound.

Since many people are getting into chickens these days, you might be surprised at some of the live chick prices from 1956: "Hatcheries reported prices paid for hatching eggs during the week at an average of 84 cents per dozen. Average price charged by hatcheries for chicks was reported at $15 per hundred. These prices compare with 84 cents and $15.25 for the previous week and with $14.50 a year ago."

You could get a "gentle mule that will work anywhere" in Blount County, and an equally able mule in Cullman County for $150. During these months Alabama's "dairy men" set a new "high mark" in efficient milk production. Milk production was 338 pounds per cow above that of 1954 and "butterfat was 10 pounds higher than the previous year."

While it may be hard for our tech-savvy youth to realize, many of the rural farms in Alabama had only received electrical services and phone services in the previous decade, and many churches' and rural schools' bathrooms were still little "houses out back." I know my own family, just three miles outside Oneonta, installed their first indoor restroom shortly before I was born in 1952, and a party line telephone was the chief means of communication. I can remember when my family bought their first TV around 1956 when these newspapers were written and when I was four.

More Simple Times at Old Field Farm

No—I wouldn't want to go back to using an outhouse (but I COULD if I had to) and there are many other modern conveniences that help me in my life, BUT I wonder many times if it really was a good trade off.

I enjoy typing an article and having it instantly in the hands of my editors miles and miles away. I'm glad I don't have to hitch the mule or horse to the wagon each time I want to run to the store as my Granndy did. BUT...from the writings of my grandmother Maud Smith Lowry in the early 1900s to these 1956 newspapers from more modern times, I wonder quite often if we've really improved our lives with many of those changes or would we be much more satisfied and CONTENT to live these more SIMPLE lifestyles....

Just something to think about from this simple woman struggling to live her simple life.

Suzy Lowry Geno

Maud Smith Lowry

Being Prepared in a Different Way

Leroy the rooster died this past week.

At the ancient age of 11 we knew his time was limited.

For the past few months, I had been catching him every night and placing him atop an empty cage in the bunny barn because he had gotten too old to get to his roost of safety outside. During the last week of his life, I carried him in each night and placed him in a clean bunny cage filled with hay for warmth.

Leroy was a special rooster (those who know me personally and all my sweet readers of this column are probably shaking their heads because they know ALL my animals are SPECIAL to me!)

But Leroy had an unusual beginning. In the early 2000s, a man regularly bought fertile eggs from me to hatch in an incubator at his farm. One week, I was lamenting to him that my broody Easter-egg hen had failed to hatch the five eggs she had been setting on for more than a month in a box on a shelf in my carport.

He returned late that afternoon with 5-day-old baby chicks. We carefully removed the eggs from underneath the hen once it became dark and substituted the day old chicks. Talk about a happy chicken! Those were HER babies and she protected them, taught them, and evidently loved them as they traveled all over this farm eating bugs, worms and grass. Leroy was one of those chicks! A fertile egg from here and hatched elsewhere, but then brought back here to lead a happy rooster life.

Most of my hens go inside their sturdy coops every night. But Leroy was raised in the yard so he never learned to go inside. My late husband Roy, always an early riser, would go outside every morning about 4 am and sit on the carport until daybreak. Leroy learned that routine very quickly!

Roy would carry a piece of loaf bread outside every morning and sit there and thump little pieces to Leroy until Leroy had eaten it all. If Roy failed to go outside, Leroy would travel from the carport to the back door and back to the carport repeatedly, crowing until somebody came outside and gave him his treat!

Roy was the one who named the rooster Leroy and the name stuck.

Leroy outlived Roy by about 18 months. And while I admit I cried when I found Leroy in his cage that morning, it was hard to mourn him when I can see his offspring running all over this farm. One of his sons is colored exactly like Leroy and is so old himself he has a few gray feathers!

Living on a homestead or a farm has made me not more comfortable about death but perhaps more assured and at peace of how death is just an extension of our lives.

Most folks who know me know I am somewhat of a "prepper." And I've even written articles for this column about the importance of having extra food, essentials such as medication and more saved in your home (and your vehicle) to be prepared to take care of yourself and your family members if something happens.

The snowstorms in January and February of this year were good examples of why you should have a "go" kit in you vehicle (just ask some of those folks who were trapped in their vehicles on the Interstate for hours because of the ice and snow!) or in your homes. Even a few energy bars, some bottled water and a warm blanket can make a huge difference.

And I've told here before of how having food stockpiled in our pantry saved the day when Roy had his two heart attacks and I couldn't leave him to go to the store and had little money when I went!

I am not a financial planner and certainly not an attorney, but maybe me telling a little of my personal story will help somebody get ready and not be in the same predicament. Most folks don't like to think about their own deaths. We kind of want to think we're going to live forever, don't we?

But unless it's the day for the Lord to come sailing through the clouds at the end of time, each and every one of us IS going to die.

If you love your family enough to stockpile food, medicine, and other essentials to have in case of an emergency, these next steps are equally important.

When my mama neared the end of her days, she thought she was pretty prepared and what she had done made a big difference. She had signed a "Living Will" or an "Advanced Healthcare Directive."

I cared for her here on the farm until about six weeks before her death. Then she was placed in a nursing home less than a three minute drive away. I fed her lunch and supper every day. When I walked in there to feed her lunch about two days before her death, nurses were bustling around her because she'd had a massive stroke. They wanted to take her to the local hospital. I asked why and they began telling me all the procedures they would be doing to her.

These were not procedures she wanted and she'd signed a legal paper saying so. A quick phone call to her doctor and she was allowed to die with dignity as she had wanted, without being moved or further prodded and poked. She died with me stroking her arm and simply stopped breathing.

There was other legal paperwork she hadn't done that would have made the next few months a lot easier. So Roy and I not only signed Advanced Healthcare Directives, we prepared new wills and prepared Powers of Attorney so each of us could act for the other in case of emergencies.

Less than two months after getting my mother's "estate" settled, Roy had two heart attacks and so began our journey of several years of him facing death as he battled a bad heart that could not be helped by an operation and then esophageal/stomach cancer.

Each hospital and doctor honored the Directives and Powers of Attorney even though we had prepared them ourselves, since we had them witnessed or notarized as required by Alabama law. Roy was also allowed to die with dignity here at our home on the farm. He had told them two months before he died that he did not want a feeding tube.

We had two more months of watching old westerns on TV before he slipped away.

I know it is better to have an attorney to prepare all the legal paperwork and that is wonderful if you can afford it. But also the Will we had prepared ourselves was honored in Blount County Probate Court because it too we had prepared according to Alabama law (look things up on the Internet and then ask a lot of questions to everyone you know who has had their wills prepared by attorneys!)

Hospices, hospitals, senior groups, and the Internet all have copies of Health Directives and Powers of Attorney you can use as your guide.

When I attended Annie's Project this past fall through the Blount County Extension Service, a wonderful farm financial planner spoke to us and answered a multitude of questions. If you have a large farm or even a small homestead and can afford the small fee to talk with someone like that, you can rest assured your wishes will be carried out legally and for the good of your family after you pass away.

I love living the simple life.

When it is my time to leave this world (which I still hope will be when I am at least 103 and die in the pen with my beloved goats!) I want my affairs to be in order so my children and grandchildren don't have to worry about last minute details.

So this week when you buy that extra can of meat (or better yet when you pressure can some of your own home-grown beef or pork!), buy those extra rolls of toilet paper for your pantry, and that extra box of ammunition for your gun safe, think about these final legal papers you need to prepare as well

Suzy Lowry Geno

You can have an extra bit of peace of mind. And peace and contentment are what the SIMPLE life is all about!!!

Leroy

More Simple Times at Old Field Farm

Christmas memories from Old Field Farm

What is it about Christmas that makes us think of the past so much!

Was it because we were younger? Is it because we miss our families? Or is it because of Christ's sacrifice that makes us want to share so much of our lives and share His Gospel with everyone?

Many of my regular column readers may know that my maternal grandparents were Vennie and Jim Inmon. Way back in 1983 I shared some of their Christmas memories with a local newspaper where I was a reporter.

Granny then remembered her childhood Christmas near Altoona in Etowah County and then in the edge of Blount County as special because she and her three brothers and two sisters looked forward to the big red apples and a sticks of candy they'd receive in their stockings. It was a special Christmas indeed when she received a small, black-haired doll whose head was made of china with a body of softly stuffed cotton.

Christmas with Grandpa Jim and their five daughters and two sons proved even more special. Grandpa was truly "Santa" and Granny his "Missus" as they sprinkled soot on Christmas tree branches so the children would see what a hard time Santa had dragging the tree down the chimney. Large sooty-black footprints also made a trail across the freshly scrubbed floor...

Granny was only 14 when they married, so looking back now I can see that they were only kids themselves as they began welcoming their family of seven. (They were married more than 50 years!)

Many of the toys their kids received were homemade, and Granny displayed a special talent for creating stuffed animals and dolls without patterns from scraps of fabric from old clothes and whatever else she could find. (Every time I make something similar to sell in my tiny general store I think of her sewing those little special critters and how my mama loved them!)

Granny ALWAYS made a chocolate cake and a coconut cake for Christmas. The coconut had to be hand scraped from a real coconut (and I remember thinking of how those coconuts---that we saw ONLY at Christmas back then---looked like little monkey faces). The coconut couldn't be from a box or bag! The children had to leave the tempting cakes untouched in the middle of the table until the big day---but on Christmas Eve, Santa always helped himself to a large piece of each! Only later did my mama and the other kids put two and two together as they remembered Granny's fondness for coconut and Grandpa Jim's for chocolate!

When the children were older, cutting the tree became a family affair, and they trooped through the woods until they found a pine that would touch the ceiling. Sweet gum balls were dipped in flour to make snowy ornaments, popcorn was strung and crepe paper twisted to produce other decorations. My mama could also remember Granny twisting crepe paper in special ways to form ropes across the ceiling that all tied together in the center of the ceiling with a special homemade ornament trimmed with twigs of holly or other greenery in the middle.

Grandpa babied Granny until his death, but a lot of good-natured kidding and joking happened through the years! Granny always "snooped" to try and determine her present under the Christmas tree. One year, Grandpa hid her real present and wrapped a brown, stuffed teddy bear and placed it under the tree.

"I punched little holes in the paper trying to see what was in that package and I got soooo mad," Granny told me in 1993. The bear survived more than 50 years and was loved by a generation of grandkids and great-grandkids, and nobody seems to remember what the "real" present was!

Grandpa was sort of a handyman and could make many interesting things. At Christmas, he decorated their yard with cutouts of Christmas figures he made and painted. Of course, there was Santa, a snowman and other imaginary friends, but there was always an angel or two and sometimes even a specially cut out and painted nativity scene. And I especially remember a little church he made and painted.

My grandpa died in 1964 but my Granny lived until the mid 1980s and EVERYONE in the family was REQUIRED to spend Christmas Day at Granny's! By the time of her death her descendants numbered near 100 and we basically all still trooped to Granny's little white clapboard house in Oneonta for Christmas dinner every year!

Although she was blind and basically homebound, she kept up with everyone's activities throughout the year and she KNEW who didn't come on Christmas, and they better have had a good excuse for missing! As long as she was able, she made certain every one of her grown children, her grandchildren, and then beginning her great-grandchildren had at least one present underneath her tree! It didn't matter to us if it came from the local five and dime store: it was personally picked out and wrapped by Granny and Christmas wasn't officially Christmas until we opened that special gift! And Christmas wasn't usually Christmas unless several gathered around the old upright piano in the dining room and sang old time Gospel songs!

I was so blessed to be a part of this huge family whose legacy continues! But the best legacy I received there and from my parents Paul and Inez Lowry and from my Grandpa Harly Lowry and his wife I never knew, Maud Smith Lowry, was the knowledge that Christmas was only the BEGINNING. That little baby boy was born in a manger to die for me and take away my sins!

The Christmas Story is a SIMPLE one. But it didn't end with the bright star above and it didn't even end with the cross...Christ arose so that any who repent, believe on Him and consecrate their lives to Him can share in His eternal life! May your Christmas be filled with Christ's true glory!

More Simple Times at Old Field Farm

(Cousins Mike Inmon, Grace Evans Huie holding Danny Inmon,

(my brother Bobby Lowry) and cousin Johnny Evans, with some of Grandpa's cutouts with a neighbor's house in the background)

Suzy Lowry Geno

Sometimes I wonder.............

Did you realize that for JUST $40 (plus the obligatory shipping and handling) you can order a scented candle that will make your house smell like you have been burning wood in a fireplace...."That sweet smell of wood smoke wafting from a chimney can make any dwelling seem homey and cozy."

That was just one of the ads I noticed in one of the sleek "country-style living" magazines the other day. And I think it said it was a teeny-tiny, eight-ounce candle! I could buy a small TRUCKLOAD of firewood for that price! (Don't worry, I didn't buy that magazine!. It was in the used magazine bin in my little general store!)

And all of this time I've been worried during the winter months on the few times I actually go anywhere off the farm because I feared my clothes, my hair, and ME always smelled of wood smoke since that's the only heat I have. Now I discover that wherever I go I'm just adding to the "cozy ambiance."

But now it's summer. It's HOT outside. The humidity has been awful so far, at least it seems so to me. And, of course, I'm living without air conditioning. The windows are thrown open wide. The ceiling fans are twirling. And I'm enjoying the sights, sounds and smells of my little homestead.

Once again, you can walk into any big store or grocery store and find shelves of deodorizers and room fresheners (most containing chemicals that you can't pronounce and probably, at least in my opinion, don't need to be breathing!). One is advertised to completely dispel the "mustiness" of your home...

I read books from years gone by and see how the old timers dispelled that mustiness when springtime approached. They opened all their doors and windows, gave everything a good scrubbing, and left those doors and windows open until frost approached! Yes. It can get hot. But I grew up in the rural South where nobody had air conditioning.

When we visited my Granny and Grandpa Inmon in Oneonta during the summer, if it was a morning visit, we sat on the squeaky glider or wobbly rockers on their front porch. If it was afternoon, we sat out back (near Grandpa's worm bed, the vegetable plot and Granny's flowers) under a huge shade tree that also shielded their coal pile in the winter.

If I visited my Grandpa Lowry "across the creek" from my early childhood home, in the summer time we sat on lawn chairs on the bank by the creek! (In the winter I sat in his lap in his rocking chair in front of his wood-burning heater, usually after we'd eaten our lunch cooked on his black square wood-burning stove!)

Later, I lived in Florida for three years with two little girls of my own. There was no air conditioning in the newly constructed house we bought there.... even my car didn't have air-conditioning! And evidently we didn't melt! We weren't used to air-conditioning so we didn't miss it.

My kids played outside so much that when my parents came down for their first visit and my oldest daughter was outside with a group of neighborhood kids, they didn't recognize her at first because her hair had been bleached by the sun and she was tan and healthy!

And there are so many added benefits of having open windows at home.

Some of you may have heard the story of my hen that laid her first egg many years ago when I was first starting with chickens. My husband told me there would be a special cackling song a hen would sing when she laid her egg.

With windows wide I was writing inside the house that August morning when I heard it! That special song! I raced outside to the chickens and sure enough, a big brown hen had laid her first and was letting the world know all about it!!! The very first egg every laid on my farm! If my windows had been closed, I would have missed that joy!

Sure, it gets hot in my house sometimes.... especially in my kitchen when I'm making jelly to sell in my little store. In my "perfect" homestead, I would have a "summer kitchen" out back where it would be cooler for such activities. BUT for now, I just do those type things early in the morning or late at night when it's cooler. Very simple.

And then there's that other thing that keeps me wondering....and if you read my columns you probably know this as well. I don't have a clothes dryer except the solar one stretched between two posts in my backyard!

Who hasn't seen the TV commercial advertising the fabric softener that makes your clothes and bedding smell like it's been hanging on the clothesline! I just hang my sheets on the line, put them back on the bed that night, and voila! I have that fresh smell and it hasn't cost me a penny!

I know everybody doesn't want to live the way I do. I know there are lots of conveniences I enjoy. But to me it seems so contradictory to work at a job away from home so you can afford all the high-costing conveniences in your home that you are SELDOM home to enjoy!

So many kids have extravagant play sets in the backyards (and even wonderful swimming pools!) but they're not around to enjoy them much because they're in after school activities or day care because both parents are having to work to pay for all those goodies.

Sometimes I wonder if the kids wouldn't be just as happy with a tire swing hanging from a tree limb and a homemade dam made at the creek at the bottom of the hill and maybe a parent or two around a little more to play WITH them...

Don't get me wrong. This is America. We can work and have just about anything we want or dream of. And it's everybody's right to do just that. Not many of us would want to go back to using outhouses and hauling water...

But sometimes I wonder.... if we have lost our moral compass as we march toward more elaborate play groups, vehicles that are computerized wonders and neatly manicured lawns around those climate-controlled houses.

Think back to your best memories a child...I bet they involve PEOPLE and not THINGS.

Lately I've had so many people tell me they wish they could live my lifestyle or they are "envious of my life here on this small farm." But they are busy counting their years until retirement. Or their health may be deteriorating Or they just need to get more bills paid off first.

It's wonderful to plan and to dream and to have hope!

But this is the only life you will have so you need to do what you really want to do NOW! You may not want to live a SIMPLE life like me, but DON'T WAIT to do whatever you want...

You can figure out how to meet your responsibilities and how to make your life the peace-filled rewarding life you want if you just try.

There's NOTHING stopping you from doing what you want to do except YOURSELF!

Have family responsibilities? Great! Nothing is better for kids than to be raised on a simple family farm. Have older parents? They'd probably enjoy that change of lifestyle too!

Have health problems? You are going to have them wherever you are so you might as well be living and enjoying where you want to be. And who knows, a less stressful lifestyle might actually improve our health!

Have a lot of debt? Pay it down as quickly as you can by doing without some of the extravagances you don't REALLY have to have....and then don't have any debt except your house payment and, once you pay off you home and farm, NEVER EVER put another mortgage on it....I am speaking from EXPERIECE...

(And just try opening those windows and shutting down that thermostat......isn't there something you want MORE that you could use that power bill money for???

So don't live your life dreaming of when you'll retire, or how things will be different when you get that raise or when you get that promotion.

Begin LIVING your life now!

Those tomorrows are never guaranteed!

Suzy Lowry Geno

Guineas.... My latest love on the homestead

That has to be the most annoying noise I've ever heard," my niece Jeanne Wood said as she sat in my yard with my grown kids.

I sat there puzzled for just a minute. There were roosters crowing, chickens cackling, goats baaaaaing, cats meowing, dogs bouncing and barking,....ahhhh....she had to be referring to the guineas!

Just like practically everything else on my homestead, I seldom notice any of the noises unless the animals are sounding some sort of alarm. But the guineas are "alarmed" at just about everything!

My journey with the guineas began in late spring. I'd heard adult guineas are bad to roam and I'd seen that myself. When my son Nathan bought his first home on the outskirts of the small town of Altoona, a male and a female guinea were included in the bargain (because nobody could catch them). He fed them and kept water outside for them, but it seemed they just cruised through their home base when they had nothing else to do!

When Nathan bought a house here on the farm about seven years later, at least one of those guineas was still roaming Altoona!

I wanted guineas that would STAY here on the farm and provide some semblance of insect and snake control. My research in Mother Earth News, Backwoods Home, Backyard Poultry and on the Internet suggested that if they were raised from small chicks they MIGHT be less apt to roam.

But I didn't want as many as most hatcheries sold in their minimum order. But then I discovered on Facebook that a woman in the Moody/Odenville area who had guineas had a few for sale! She graciously delivered the 10 little peepers to me because she wanted to see my farm's little general store.

I had thought they were going to be a little bigger. Although they were fully feathered, they were still too small to go in the isolation room in my chicken sheds. So of course they spent their first three weeks here in a giant washing machine box under a heat lamp on my back porch. (If I had a dollar for every critter raised in boxes on that back porch, I'd be sitting pretty!)

The first night they chattered so weirdly that Nathan said all the way to his house they sounded like mini-car alarms! But after a few nights they quieted..... some....

I then kept them in the isolation room in the chicken sheds (a large area where they could see the other chickens come and go, but where no other birds could peck whoever is currently in the room).

After three more weeks, one morning I cautiously opened the door to allow them to free range in the large fenced area on that side of the shed. After a little apprehension, they sauntered out and then chattered at everything that moved! By the next week, I was allowing them free range of the entire farm!

The first couple of nights, most of them went back inside their room for me to safely shut them in.

But, by the end of that first week, all 10 of them were roosting each night in the tall pear tree that stands above my front goat pen. (That's where the hatched-on-the-farm-roosters and a couple of game chickens also roost). They are generally safe there and they've roosted there every night since!

They are great alarms! Evidently they can hear or sense coyotes coming through about five minutes before I hear the yipping and yapping, so I have time to grab whichever flashlight currently has a charged battery and slip on some shoes before the coyotes race through the pasture.

Guineas are also known to be great at insect control, especially ticks; they are often brought in to areas plagued by Lyme disease. I really believe they have helped with snake control since I've not seem to have as much of a problem with snakes in and around the farm, especially the bunny barn, since they have been patrolling the area.

There are mixed ideas about guineas and snakes from the "experts." Some say young guinea keets will just gather around a snake, sounding the alarm with a special kind of chattering. Others say older guineas will kill even large snakes, and there are some pretty graphic videos on the Internet to supposedly prove that.

Probably the biggest expert on these weird fowl that likely originated in Africa is Jeannette Ferguson. She bought guineas about two decades ago, to rid her garden of Japanese beetles and grasshoppers, because she wanted beautiful flowers to enter into her area's garden club competitions. Evidently they worked for her because she's since won more than 100 prizes for her lively flowers! And she also noticed the side benefit of finding fewer ticks now---even though she lives in an area prone to tick infestations. (And, guess what, they are said to LOVE to eat fire ants!!!)

Since Ferguson had trouble finding information on guineas when she first began researching them, she wrote her own book, "Gardening with Guineas."

You are supposed to be able to tell males from females by the wattles on their necks and their different way of chattering their call that sounds something like "buckwheat, buckwheat." But I haven't been able to tell the difference in any of mine.

We added a few more guineas this past year unexpectedly when a customer at my little store showed up one morning with a five gallon bucket that appeared to be filled with straw. She asked didn't we have an incubator and I told her that my son Nathan did.

She parted the straw and there were about 17 guinea eggs in the bucket. She said her guinea had just begun setting that day and something had gotten her during the night. Nathan took the eggs and placed them in his incubator expecting about a month long wait....less than two weeks later he discovered just-hatched guineas running around inside the incubator!

Evidently that guinea mama had been setting longer than my customer suspected! And 15 of the eager little things had hatched! They didn't come out of the eggs slow and dreamy looking like baby chicks. They came out ready to go!

I brooded them under a heat lamp in a big pasteboard box on my back porch and they all grew to adulthood. We've lost some since then to the highway and one simply dropped dead in the barn yard. But there's plenty still running around!!!

They still travel mainly in a pack. If you see one, you'll likely see all of them. There was one slightly smaller that was always bringing up the rear in the original ones and I surmised she might be a female. But, as they've grown older now, they're all about the same size.

There is one who acts slightly lame. When I finally let the younger hatchlings out to free range, that one lame guinea seemed to always be watching out for the younger ones, so I do suspect she is a female! And by the way, she doesn't let her "disability" hinder her in any way, and can still fly with the best of them as well!

Guinea eggs are edible and some folks love them. Usually they don't start laying until the next spring after they are hatched. They don't make very good mothers (too busy chattering and nosing into whatever is going on all around them), so if you want to hatch out some yourself you either need to gather the eggs and hatch them in an incubator or under a brooding regular chicken. But like everything else there are exceptions to that rule and one of my friends had a guinea mother hen who hatched out her own brood this past spring!

While I love my guineas, I'm afraid that if my late husband Roy was still alive, the guineas would have long been gone! Their noise throughout the night is almost constant and sometimes I wonder if they ever actually sleep!!!

But I am so used to their signs that unless they are really making the "something is badly wrong in the barnyard" sounds, I can tune them out just like I do the many rooster crows!

So if you don't mind their noise (and you don't have close neighbors who might object!), I'd recommend my noisy friends to every household!

Suzy Lowry Geno

You aren't here..............

This morning I made blackberry jelly at 5 a.m. Before the heat encircled my little kitchen—but you weren't here to help me eat up the excess on toast while the little glass jars were water bathing...

Blount County's clear seed peaches are ripe! But you aren't here to eat that first whole basket full with juice dripping down your chin....

The swallows are back with their messy nests! Jannea is making sure they are safe and that all the babies get equal feed from the ever-busy parents! But you won't be here to watch delightedly as they fledge---first flying to the edge of your ceiling fan on the carport patio and then racing out into the unknown...

You're going to once again miss Blount's first ripe cantaloupes and you won't be talking one of your farmer friends into letting you have a big batch of plain old time field corn for me to cook for hours with flour and salt...

You missed Shadow-Big Puppy's arrival on the first week of June last year! How he moved in here after none of the rescue groups could catch him and treat his many ailments. One lop-sided grin on his big, old Great Pyrenees head and I know he would have captured your heart as well!

You've missed Nathan bending over the blueprint table in his office, pondering electrical jobs that I know he still wishes he could ask your advice about... and you missed the birth his baby, Aria.

You've missed Kori and Arthur having their own little Old White Oak Farm complete with chickens for Devin, Savannah, Kelsea and even Kayla to sometimes tend...

You weren't here in April when great-nephew Peter was commissioned a Second Lieutenant!

And there was my very first book---a collection of columns from these Simple Times articles! What a dream come true in my little world! But you were a part of it; the photo your Granny made of you when you were 6 years old standing barefooted in a Morgan County cotton field is on the cover of that first book. But you weren't here to share my joy when that first printed copy was placed in my hands by my publisher.

You weren't here to fuss when I got noisy guineas! And you weren't here to complain about the 31 smelly chicks in the big pasteboard box brooder on the back porch. And you've missed the farm's first turkeys and all the antics they've created!

I wish I could tell you that we're all doing fine, because we are.

But it's still sad that there have just been so many things and times where "you aren't here...."

The Center for Disease Control has statistics showing that 442,000 other folks in the United States died in the same year, 2012, as you, directly from smoking.

Dr. Miachel Roizen, an international medical specialist often quoted on the Internet, says there were four million deaths from cigarettes worldwide, making it "the leading cause of death in the world, higher than infectious disease, greater than obesity, greater than guns."

And then there's Dr. Mehmet Oz, cardiologist and TV star, who simply states that cigarettes "are one of the most preventable causes of premature death."

I still thank God every day I was able to keep you home here on the farm during your illnesses and you died peacefully here at home. That was indeed a blessing. But with you being about 6 feet 5 inches tall and maybe weighing about 60 pounds when you died, it was not an easy illness or an easy death.

You fought bravely back from two heart attacks that damaged your heart so badly there was no surgery, no stent, nothing the cardiologists could really do to help. But you didn't quit smoking.

Then you were gripped with esophagus and stomach cancer, COPD, deep vein blood clots and so much more.... but you never stopped smoking...

Just this last month I had a major allergic reaction to poison oak. Somehow it affected my breathing and my throat and I had to rush to the emergency room gasping for air.

For two weeks it hurt every time I breathed and I'd wake up at night gasping and wondering if each breath might be my last. I was sicker than I have been in years. And I thought of you.

Those last few months you were so short of breath. There were inhalers and ventilators and breathing treatments. Eventually there was oxygen part time and then full time.... You experienced months of having breathing problems while I only experienced a few days. But I'm nearly well now. And you're not here...

More Simple Times at Old Field Farm

I ran into your old golfing buddy John in the grocery store. He hesitated, but then asked me how your cat Sadie was doing.

Sometimes I wonder just what Sadie is thinking. She'll look over to where your recliner sat in the corner and she gets a faraway look in her eyes...maybe I'm imagining. But I'm wondering if she's thinking of things where "you aren't here" as well...

Suzy Lowry Geno

    I'm sorry you're not here to enjoy the wonderful new church I'm now a part of.... the church just began around 2012 I think, right about the time you were leaving us...You'd know many of the folks that are there. It's a Bible-based Gospel-based church like we were seeking.

There are a lot of wonderful young families. But I'm one of a couple of widows there. That's not a part I wanted to play at age 60 and I'm still not enjoying that part at age 63.

The first week in August of this year you will have been gone three years. You left us August 6, 2012, the afternoon before your 68th birthday.

Instead of a Bible verse on the program at your funeral, I had them print this simple message: "If you smoke, please stop. If you don't smoke, please don't start." That's also my wish for everybody now reading this book.

I really wish I could let you know how well that I am doing I am blessed. I am happy.

My little farm, my animals, my little general store, my life are all doing fine.

I have true peace in my daily simple life.

But you aren't here...

Suzy Lowry Geno

Wood heat warms your soul

Just about every article or TV segment on "saving energy" begins with the statement, "Turn your thermostat down (or up depending on the season) to so-and-so."

That would be well and good if you HAVE a thermostat, but there are more SIMPLE-living folks than you can imagine who manage to live 365 days a year without that little magic dial or digitally-glowing gadget on their walls.

Many folks told me last summer that I had lost my mind because I have lived through the Alabama-heated summer yet again without an air conditioner in my house. (Remember, Southern girls don't perspire---they glow"!)

And there's even more folks who question my sanity when they learn my sprawling house is heated solely with a wood-burning heater. But I just smile to myself because I know something too many people don't know: wood heat warms your very soul!

Some of you may know that on Halloween (2013) I slipped in a very strange way and even though I never hit the floor, I sustained a severe hamstring injury. Yow! It was (and continues to be) one of the most painful things I've ever gone through (I have new sympathy for football players who get that injury!!!)

My youngest daughter was being helpful and brought in a so-called energy-efficient room electric heater so I wouldn't have to carry firewood in and get up and down to stoke the heater. I used it for three days and while it seemed to keep the room warm, I nearly froze to death!

Yes, it could have been in my mind, but when I rebuilt the fire in my sweet wood burner, the house took on its usual warmth. And I could stand up against it or sit nearby and just feel the warmth reaching inside to my very bones.

You can't do that very well over a vent with a forced-air heater, can you?

I know wood heat isn't for everybody but here's how it became such a big part of my life.

When I was a little girl and we lived just down the hill from my current farm (one of my allergy doctors wrote that I lived "in a hollow beside a creek"), we heated with a propane heater in the living room. But my parents' bedroom where my little bed was also located, had a fireplace.

On really cold nights, they built a fire, sometimes with coal and sometimes with wood, and I'd go to sleep comforted watching those flames and hearing the wonderful crackling of the fire...

Many years later, my husband Roy and I bought our then-dream house. Totally electric, central heat and air. Just touch that dial. We touched. Then we flinched when the bill came!

We traveled over the river and through the woods to Cullman County and came back with an Ashley wood-burning heater, a three-wall safety pipe to go through the ceiling and roof, and a fireproof pad for the floor. We were in business and we loved it! Roy was still young and healthy, and he and my dad cut mountains of free firewood from my home place.

Then our house burned!

But NOT from the wood heater! The cause was electrical wiring behind the clothes dryer. The wood heater was fine, so I sanded it down, painted it myself with the right kind of paint and we installed it in the new house we built back. We even built it safer.

There were special tiles laid on a cement pad on that section of the floor and up the wall behind the heater. Again we had three-wall pipe through the ceiling and roof. It was even easy to clean out the chimney because I could take out the elbow by loosening a few screws, and clean it out from the bottom up with a large box situated underneath to catch the soot and creosote.

Then my daddy died and the following year we bought the house and farm from my mama. My precious wood heater was solder with our other house...

We managed for a couple of years without it until the Blizzard of 1993 hit. Have you ever tried to heat a sprawling house with just a fireplace for six days??? We slept on a mattress in front of the fireplace and vowed to never again be without a wood-burning heater!

I bought another Ashley from an Oneonta business and we set it on the raised hearth of the fireplace. Putman's Tin in Oneonta cut a piece of metal to fit the fireplace opening with a hole for the pipe But instead of just venting the heater up the chimney, we have pipe that goes from the heater up through the brick chimney to the top---just an added safety feature.

It's a major pain to move the heater out to change the elbow every three or four years, but that's what a strong son and friends are for! And I can run a brush up through the heater and up the pipe to clean out soot and creosote more often.

A couple of years after we got this wood heater in the 1990s, our central heat and air unit died completely. We never replaced it.

After Roy's heart attacks, all heater duties fell to me. After he passed away, tending the heater was no big deal because I'd done it for so many years after his sickness (and years before when I worked at home while he worked running his electrical company).

Last year I even borrowed a neighbor's small electric chainsaw and made quick work of several smaller trees that needed to come down around my little general store and around the farm. You have to remember SAFETY FIRST when dealing with any kind of chainsaw, so please, please, please read up on all that before you ever pick up a saw!!!

The Clemson Extension Service has some wonderful information on line at the National Ag Safety Database www.nasdonline.org.

They note." Most people have some way to arrive at the amount of conventional fuel it takes to heat the house for an average year." They quote the "Smithers" method that "assumes the following equivalents to one cord of average dry hardwood. Each of these is supposed to equal a cord of wood: 150 gallons of No. 2 fuel; 230 gallons of LP gas; 21,000 cubic feet natural gas or 6.158 kwh of electricity.

They have further info there to figure out what your savings could be. There's also helpful info on pricing a wood heater to suit your specific needs.

Several websites talk about the release of pollutants from wood heaters and I believe contrary to what the EPA is trying to state it equals out. Numerous websites I visited such as www.woodheat.org notes, "A wood fire does not contribute to global warming because no more carbon dioxide is released than the natural forest would release if left untouched. Using wood for heat means less fossil fuels burned, less greenhouse gas emissions and a healthier environment." While I don't personally believe in the so-called Global Warming, to me they make a valid point.

You should just be careful and not burn treated wood plastics, colored print sections from the newspaper...anything that might contain chemicals.

For the last few years, I stored most of my firewood on the carport instead of under my long back porch. (I saw plans for a simple wood shed like my Grandpa Lowry used to have but the only logical place for it would be rite under the electrical lines coming into my house, which is a major no-no.

By storing the firewood on the carport, I have about a six-inch step to come inside and that's it. That means a lot as I'm not a spring chicken any more and those eight steps up and down off the back porch to get firewood from under it are a little risker with a big armload of firewood. Also with it on the carport I don't have to go out into the weather to bring in the wood and I just bring in what I need right then. At night a little extra wood is stored in a large plastic container near (but not too near!) the heater. I did store it in one of those half-whiskey barrels, but alas, it finally fell apart---and then I burned it!)

The majority of that firewood is gone by spring, so insects AND SNAKES aren't a problem. Seasoned wood can still be stored under the porch until I'm near needing it, then it can be moved to the more convenient carport.

As with any type of heat, I always have at least four smoke detectors with FRESH batteries in areas of my house. It might be a little easier to flip a switch or turn a dial for heat, but there is no describing how it feels to sit in a rocking chair in front of the wood heater, sewing a quilt or knitting, while my cats curl around my feet!

Suzy Lowry Geno

(This article on wood heat was the first place winter in the National Council of Farmer Cooperatives contest for 2015)

Rabbits on the homestead

I was instantly in love.

It was the mid-1990s. There on the cover of my Country Woman magazine was a woman holding what appeared to be a giant cotton ball with upright ears and two beautiful eyes. I quickly tore open the magazine to the short article inside to learn the amazing looking creature was an Angora rabbit---and my life was changed forever!

Here was some livestock I could get serious about!

I couldn't imagine sharing a full-size sheep by myself. But these small bundles of wool could be shared with simple scissors while they sat docilely in your lap. I began asking around (This was shortly before everyone was hooked to the Internet where just about everything imaginable is available at the click of a button!)

There weren't that many in north central Alabama, but I soon located EIGHT for sale at a petting zoo which had decided the small beasts were just too labor intensive for their farm. I bought the rabbits complete with cages with special nesting boxes that hung lower than the actual cage and I was in Angora rabbit heaven!

My son Nathan came in shortly afterwards and asked why there were Ewoks in the backyard. They did look a bit unusual!

But whether you share my love for fiber or just want a general all-purpose animal for your farm or even your city lot, rabbits might be your answer!

There are approximately 50 breeds of rabbits recognized by the American Rabbit Breeders Association. If you want a general purpose rabbit to provide meat for your family and wonderful manure that can be placed directly on your garden or flower beds WITHOUT having to compost it, then you'll likely want what is called a "meat" breed. (I'll quickly add that I wouldn't ever be able to eat any of my rabbits. So they have to "earn their keep" with their manure and fiber!)

If what you want is mainly pets (or are just experimenting to see how you or your children will do in caring for small livestock) just about any rabbit breed will do.

If you do want to consider a "fiber" animal, you can check out the four main types of Angora rabbits. A weal of information is available on the Internet. Just type in "rabbits" in any search engine and you're on your way. My rabbit bibles to begin with were Bob Bennett's "Raising Rabbits the Modern Way" and the Northern California Angora Guild's "Angora Handbook."

There are even forums on several homesteading sites on the Internet now where you can ask questions and glean all kinds of rabbit know-how. My actual "rabbit raising" began long before my life with Angoras.

When my children were little, my parents kept a couple of huge white rabbits as pets for the grandchildren. They would actually put tiny harnesses on them and my kids would walk them around the farm on long leashes!

When Nathan was about 10 he threw a ring around a small bottle at the Blount County Fair and won a tiny brown rabbit! Now I don't condone rabbits being given away as prizes and PLEASE don't buy baby chicks and baby rabbits as Easter presents because they are usually doomed to quick and sudden deaths. BUT that little brown rabbit that Nathan won lived a long and happy life!

Nathan named the rabbit Jack and, to begin with, he had the free run of our large screened-in back porch. He quickly learned to use a littler box (house rabbits can be trained easily!). And he lived happily there for several weeks.

Then one day Nathan was near hysterics. Jack had climbed the long pipe into my then-clothes dryer (this was in the early 1990s when I still HAD a clothes dryer!) and had cut off most of his nose!

Thinking Jack was doomed, we rushed him to the local vet. After some repairs, Jack came home to live many more happy years, although he never again had a proper rabbit nose.

Rabbits are bad at cribbing. They will chew on just about anything wood. If you try letting a rabbit have the run of your house, they will also chew on electric cords, which can have disastrous results. If you're going to have a house rabbit, it's best to keep it contained in a cage except when you can be with it to watch it roam and run.

Jack moved to a metal cage outside. He later sired several generations of offspring and lived to be a healthy old age, dying in Nathan's arms many years later.

So I had some rabbit experience before getting my original Angoras.

Whatever kind of rabbit you decide on, a metal cage is best. Wood can harbor disease and is a haven for ear mites!

Chicken wire can be easily chewed through by rabbits and rabbits running free don't last long thanks to coyotes, hawks and speeding cars!

Jack and our brown lop Tadpole lived for a time under our back porch in their wire cages sitting on bricks so their cages were off the ground. Eventually my husband Roy helped me build an addition on the back of another barn building, which we now refer to as the bunny barn.

I hung cages from the back wall and from the front. Hanging them from the back wall was not a good idea because the wood did harbor dampness and other things. The rabbits hung near the front that was just hanging from the ceiling on small chains did better so eventually I tried to change all the cages away from the walls. At one time, I had as many as 35 Angoras housed that way!

I do not like watering system where the water goes to each cage in small tubes. I prefer individual water bottles so I can see if a rabbit is not drinking enough, and, if a rabbit is to be medicated, I can easily place the medicine in the rabbit's individual bottle.

I usually have at least a couple of "rescue" rabbits in the barn as well. These are usually rabbits someone bought for their kids or grandchildren as pets and then realized they were messy or just not good fits with their family. Some rabbits like to be petted and held. Others don't. It's just their individual preferences.

Tadpole was a brown lop I bought at a local flea market just because I thought she was cute with her lop ears. The seller "guaranteed" that she was a buck...Tadpole proved him wrong when she built a nest of hair and fur and had seven half-Angora, half-lop babies about five weeks later!!!

Since I do not show my rabbits and do not generally sell the babies, I didn't worry about babies who weren't purebred. I was excited to see the results of the breeding between the white buck and the brown lop. I wound up with two brown, two black, two gray and one kind of spotted!

Some of the kits (that's what baby rabbits are technically called) had upright ears like the English Angoras and some had the lop ears. A couple had one ear that stood straight up and another that hung low like the lops!

Since I was mainly interested in their wool or fiber, I began breeding to get different colors. As for breeding, there are a few simple things to remember. ALWAYS take the doe (the female rabbit) to the buck's cage because females are very territorial and love can't happen if the doe is beating the buck up!

Watching a doe build her nest is amazing and is a fantastic learning experience for youngsters. The doe will take hay and hair she pulls from under her chain area, and build a warm, soft, cozy nest to have her babies.

My rabbits always did better having kits in the winter, even on the coldest days, than they do in the summer.

Rabbits do not generally do well in the heat, especially long-haired bunnies. But any bunny most be kept in the shade in the summer. I also freeze two liter drink bottles and alternate them each day of the summer and the rabbits lounge around on the cool treats. I keep two box fans blowing on low, but NOT blowing toward the rabbits.

In the winter, I make sure the rabbits are kept out of the wind and make sure they have hay in their cage to burrow down in.

Rabbits are generally inexpensive to keep, and to me are fund and rewarding. I am still not progressing on learning to spin on my Ashford Traditional spinning wheel as fast as I'd like to be but I'm continuing to press forward. Sometimes it's just hard to do all the projects I want and need to do on this little farm!

I've added Jersey Woolies to my Angoras now and hope to begin breeding for different colors again soon.

If you have children, especially young children, teach them to never ever pull a rabbit's ears or pick them up by the ears. Also, as we mentioned, some rabbits like to be held and petted, and some don't. But most all enjoy being fed an occasional treat.

In all my years of rabbit raising, I've only had two biters. And both of these came from other places where they didn't have the best of care and so I believe the biting was developed as a self-defense measure. Wearing big gloves helps me when I had to handle them.

Wearing long sleeves is usually wise whenever you are moving rabbits from cage to cage, cutting their hair, etc. because rabbits can scratch you badly, especially with their back clawed-feet, even if the rabbits are gentle. They are afraid of loosing their balance and sometimes scratch when they are not meaning to hurt you!

If you are trying to develop a more self-sufficient homestead or farm, there are lots of tutorials and information on rabbits on the Internet.

If the thought of butchering and eating such a cuddly critter is not to your liking, rabbits can still pay their way with their great manure! I have grown some of the biggest cabbages and juiciest tomatoes using rabbit "fertilizer!"

So I guess you might add another title to the simple, gray-haired homesteader's names: in addition to being called the "Chicken Lady" and the "Goat Woman," I guess you can call me the "Rabbit Woman" now as well!

More Simple Times at Old Field Farm

Suzy Lowry Geno

High tech or low tech,
we'll still all find a way to keep in touch............

This morning was like most mornings. I fed and watered all the animals, showered, then grabbed breakfast and turned on the computer to check my email and messages on Facebook.

That's much different than two or three decades ago!

Then were completed morning chores while listening to local radio. And when I was growing up, my breakfast was ALWAYS accompanied by local radio WCRL streaming from mama's little beige and turquoise plastic radio on the kitchen counter.

Then local radio broadcast who died (and what their funeral arrangements were) (and usually even whose mama, grandpa or grandson that person happened to be and maybe even what caused their death!), who had a birthday that day AND the next day, who was admitted to the local hospital and who'd had a baby during the night!

Now local radio still broadcasts the deaths and on into the hour birthdays after "celebrity" birthdays, but the radio station would be sued beyond belief if they reported now about hospitalizations and more!

Another great source of local news just a few years ago in rural areas were the local community columns in the weekly newspapers.

We laugh now when we look back and see where "Frances Smith got a new refrigerator last week" or "Vennie Inmon received a LONG DISTANCE phone call from her daughter in Michigan!" Seems like nothing to us now but if you were living in those times you'd rejoice with Frances because the rural electric lines hadn't been extended to her house where she could get a refrigerator until that spring. AND the Inmons were scared (that it might be bad news) but delighted if it was good news when the telephone brought a long distance call from the daughter in Michigan or the one in Fayette because the usual mode of communication for them back then were penny postcards on which my Granny wrote with a stubby yellow pencil!

And what about reading about the Johnson family "motoring to the Gulf Coast" last week! Before interstates and more reliable cars and tires, that was a major vacation and everybody was envious but they shared in the joy as they read many of the details of that beach trip in their local newspaper!

The Prickett's dairy's unusual large quantity of milk that week, who brought in the first red ripe tomato to the newspaper office or who later brought in the largest pumpkin or watermelon (with photos duly recording the events) were big news that almost everybody liked to see or read about!

My birth even made the FRONT page of the local newspaper when I was born in 1952. Somewhere here there is a clipping showing that "Paul and Inez Lowry were the proud parents of a new baby girl" born at one of the tiny local hospitals that itself is no longer in existence.

I also have a yellowed clipping of when my great-grandfather, A.S. Lowry died. It was also on the front page complete with photo and doesn't sound anything like the obituaries of today:

"On March 1, 1922, the death angel visited our midst and claimed the spirit of Uncle Asbury Lowry. He was born in Etowah County January 27, 1851 and moved to Blount County when he was 19 years old"

"For more than two years he had been in declining health but hopefully and cheerfully he made a brave but losing fight against disease and finally the tired, suffering body fell into decay while the spirit returned to God who gave it."

"Uncle Az, as many were pleased to call him, was a good citizen, companion, father and neighbor and indeed one of God's noblemen---both loyal to God and his church, having been a member of the Missionary Baptist Church for 50 years."

"He was honored by his brethren and trusted by all who knew him: Relatives and friends mourn his passing away, but since he lived and died in the swelling triumph of faith, we press on up the glorious path..."

The obituary goes on to tell of him marrying my Great-Grandma at the age of 24 and lists their nine children, one of whom was my grandpa Harley.

It also states that his funeral was held at his home, three miles east of Oneonta, very near the place my little homestead now sits!

You don't read too many obituaries like that these days! But those old obituaries contained news about the person who died and their families.

Asberry A.S. Lowry

Nearly three decades after Great-Grandpa Asberry died, the community, and many others liked it, had a new way of keeping up with the news: the party-line telephone!

Even when I was a teenager in the 1960s we still shared the phone lines with eight other families! It seemed each community had at least one person---usually a lady who had too much time on her hands---who "listened in" to everybody else's business. Then she took it upon herself to spread the news---good or bad---to everybody she came in contact with!

My mama and daddy applied for and received a private phone line in the late 1960s after a then-boyfriend and I made up outrageous stories and talked about them on the phone knowing they would be broadcast throughout the community!

The very first phones (right before my time) were even more suited for "news broadcasting" as all calls went through a central switchboard in Oneonta manned by a couple of ladies working shifts. (Think Lily Tomlin on the old Laugh In TV show of the 1960s, "one ringy-dingy, two ringy-dingies). Everybody's phone rang when anyone received a call, maybe one ring for you or two for your neighbor and you were SUPPOSED to only pick up when you heard your particular ring.

But if you knew a neighbor was in ill health, expecting a baby, or if they were expecting some other important news, like news of someone away in the Service, everybody else on the line would wait a second then quietly lift their receivers to hear the news as well.

And if it was really important news, like someone had been killed in the war or someone's house was on fire, the Central Operator might ring up EVERYBODY and tell them what had happened so they could help their neighbors

An old hymn book includes an old Gospel song that youngsters used to cell phones and instant messages wouldn't even understand. It's called the Royal Telephone, with words and music by F.M. Lehman, and notes: "Central's never busy, always on the line. You may hear from heaven, almost any time...There will be no charges, Telephone is free, it was built for service just for you and me. There will be no waiting. On this royal line. Telephone to glory always answers just in time!"

So when you're on Facebook or other social media and lamenting that you don't want to read what somebody cooked for dinner that night or where they went that afternoon, just bear with them.

More Simple Times at Old Field Farm

I can hear from my great-nephew who is a 2nd Lt. In the Army instantly no matter where he is in the world. Likewise I can keep up with my kids and grandkids (and my GREAT-grandson) even though there may be many miles between us!

I do think we may have lost something from communication and news sharing in the "old days" but knowing instantly when I need to pray for someone or rejoicing the MINUTE they come out of a successful surgery kind of balances that scale!

And that's why practically everyone who's striving to lead a very simple life, even those completely off the grid, usually first thing hook up a solar panel or some other alternate power source, so they can connect to the Internet!

Suzy Lowry Geno

Wired and unwired

Yesterday in between rain storms, I searched throughout my farm's fields for dandelion blooms so I could begin making dandelion jelly.

This morning my chores included feeding goats, chickens, ducks, guineas, turkeys, and rabbits, and then gathering eggs before carrying in the seven armloads of firewood my wood-burning heater will consume before this time tomorrow.

There's some early sage drying in the dehydrator.

And I'm hoping to have a little practice on my spinning wheel using a combination of Angora rabbit fiber and sheep wool (from one of writer Sue Weaver's beautiful animals) before the day is through.

This could have been a typical day on this land in the 1930s and 1940s when the Granny I never got to know was roaming these hills, gathering eggs and making butter.

But there's one big difference and it's been exemplified again this week.

I'm completing this article shortly and then will send it wirelessly through the internet to my bosses Jim and Joyce who I have only met face-to-face once or twice in the nearly decade I've been writing for them.

The words never touch paper until they are printed in the paper magazine copies of the Co-op News.

Likewise the photos, which would have required me or someone else to spend hours in a smelly dark room a couple of decades ago, are now taken with digital cameras, down or uploaded to my computer, and sent across those same airwaves, touching paper only later when the article goes to press.

You're probably thinking, well so what? Most folks utilize the internet these days. And that's true. Whether it's on a desk top computer, a lap top, tablet, or a cell phone, the majority of folks spend their days tethered to the world wide web in one way or another. The person who is NOT connected is usually the exception to the rule at this point.

BUT I believe in the simple life...and so do many other back-to-the-landers homesteading types who are swelling the countryside once again and wanting to live far away from the beaten path. But this is different from any of the back to the land movements of times past. In the 1930s and 40s folks who decided to try the simple live and moved far into the woods seldom even had a lined-telephone, not less access to everybody in just about the entire world!

And think about Henry David Thoreau during his experimental living time at Waldon Pond during the mid 1800s. The solitude and silence were the two things he treasured.

But if you look at Social Media, there's whole "groups" who are "off the grid" or call themselves "simple solar" or are involved in "survival living" full time, yet their computers are right there with them in the wilderness! That computer may be powered by the sun or even wind power but it still connects them to the world.

I can remember when folks who lived remotely often turned to HAM radios to have some contact with the outside world, even if it were just in case of medical emergencies, and that hasn't been but about three decades ago.

Two happenings made me start thinking along these lines...My homestead is completely "wireless" now (yes I know I'm behind the times but it's still amazing to me!)....so I can be out in my tiny general store or snuggled in bed and just keep on working (or keep on learning!)

According to the U.S. Department of Agriculture estimate in 2013, 70% of U.S. Farms have access to a computer but only 40% of farmers overall reported using the internet for business. But those statistics change according to the age of the farmers. According to the American Farm Bureaus 2010 Young Farmers and Ranchers Survey, nearly 99% of all farmers and ranchers between the ages of 18 to 35 have access to and use the Internet and nearly three-fourths of those in that age group surveyed had a Facebook page.

And income has some bearings on usage as well. Another 2010 USDA report showed that more than 7-% of farms with sales of more than $250,000 annually use the internet for farm business.

USDA says farmers use the internet for everything from marketing crops, increasing work flow, ordering equipment, utilizing GPS when planting or fertilizing crops, or keeping current with regulations.

More Simple Times at Old Field Farm

Time don't matter to a pig....

Ever since I was a tiny girl, I've read everything I could get my hands on. The written word just shouts out to me. Whether it was the back of the Pop Tart box while I was eating breakfast or even the soap wrappers while I took a bath!

Way back then, I read just about every book in the children's section of our then small public library and was allowed, with close oversight, to check out books from the adult section. That's how I read all the Perry Mason books when I was just 11!

So just because I was recently reading a magazine article about starting a hog farm does not mean there will be porkers oinking around my homestead any time soon. But the article had some wonderful insights.

The farmer interviewed was talking about his choice of heirloom pigs, the kind that many of our great-grandfathers or grandfathers raised. He talked of the better meat, better disposition of the animals and more.

He also talked of their drawbacks as compared with raising the more commercial-type hogs of today.

He noted it usually took the heirloom pigs a few months longer to get to eatin' size. His consensus was profound, although maybe not grammatically correct, "But time don't matter to a pig!"

As I thought about the last hog farm I visited, where all the brown and white free-range pigs were lounging in a pine thicket in the shade, I wondered if those pigs might have figured out much more than us humans!

When my husband lay dying nearly four years ago, I only heard him once express any regret about his life.

When talking to a pastor who had come to visit I heard him say simply, "I thought I'd have more time..."

When I was just a child, Disney and other TV shows about the future showed how we would have by now become a society or leisure because of all the time-saving inventions that would help us in our jobs and our homes. Even the cartoon "The Jetsons," showed robots doing all the housework from cooking to cleaning.

From microwaves to computers, we are now living in that future, but statistics show folks are working way more hours at their money-making jobs and housework is often still overwhelming despite being able to zap food, do laundry at the push of a button and even vacuum our homes with little round robots that work on their own.

Why has our quality of life not improved? Why are even more folks depending on pills to quiet their anxiety as they rush through life? Have we lost our focus?

Someone brought me a couple of books by Amish-farmer David Kline, collections of essays or articles he had written and first published in an Amish magazine.

While I don't agree with all the religious aspects of the Amish, I think we can learn a lot from many of their ways.

He tells of how they all ran out of the barn when his children hurried in to announce that geese were flying over. As they all raced out into the snow, they discovered not geese but eight tundra swans.

This was extremely exciting to them as they had never seen swans on their farm before and these swans became No. 136 on the list of birds that he and his wife had listed since their marriage!

In other essays he tells of how, as he walked or sat behind a team of plowing or working horses, he was able to share all sorts of other things in nature with his children. They all worked together on the farm so they were all able to experience the joys and sorrows of day-to-day life together.

There was no hurried commute to a job in the city. And no blaring TV, as they ate a fulfilling breakfast around the table after early barn chores were completed.

Kline quotes Wes Jackson in talking about how horses restrict the Amish farm work in a good way, "...but horses are ideally suited to family life. With horses you unhitch at noon to water and feed the teams and then the family eats what we still call dinner.

While the teams rest there is usually time for a short nap. And, because God didn't create the horse with headlights, we don't work nights."

He notes, "Probably the greatest difference between Amish farming and agribusiness is the supportive community life we have."

While that sounds leisurely, he writes about Amish farms that are diversified, not just raising one primary crop, and how every day, every month, specific tasks must be completed. He also details about a neighbor needing help bringing in the hay (because of a leg injury) or when all of his immediate family were shocking the wheat trying to beat a late afternoon thunderstorm. When they got to the top of a small rise, they saw their neighbor's family shocking toward them from the other end of the field! Just coming out to help because they knew a storm was coming!

When the fieldwork was finished, they spent time visiting and eating homemade ice cream!

They don't have to worry about huge farm debt for machinery, because they generally pay for their horse-drawn equipment as they buy it.

They don't have to worry about huge farm debt for additional land, because most farms were from 20 to 200 acres, according to the size of the family and the crops they plan to grow.

Even the more modern Amish who now make furniture in barns that were once filled with livestock, still practice these same values...

So many times today even folks who aren't farmers don't see their families much. They are busy commuting to their jobs in another city, paying for a huge house they only see in the darkness as they drive in from another day at the office. Working overtime for homes they don't really get to enjoy.

When I was a youngster, there were kids playing all over the neighborhood. Now, too many times, kids are well-cared-for and enjoying supervised play at daycares....But I still think there is a lot be said for standing barefoot in swirling waters and damming up a small creek with fistfuls of mud with your slightly older cousin...

When I was little the only families who had second vehicles were those whose "extra" vehicle was a rickety farm truck! There were no bills for cell phones, no money needed for fancy nail jobs (how in this world would you mail a cow or a goat with those things on!) and a family very seldom ever ate in a restaurant. So what little money we had went for the basic bills...

Noted Pastor John MacArthur talked about how we should interpret Ephesians 5:15 and other verses: "Each of us are presented every morning with another 24 hours and it is up to us to decide how we are going to use that time...The time to make the best use of time is now, not later...."

Someone asked one of those questions on Facebook this past week: "If you knew this would be your last day on Earth, how would you spend that day?"

Think about it....

The piggies at Piggy Patch Farm in Blount County don't have to worry about time (or anything else!!!)

Made in the USA?????

"Humpfgh....." the man made a half-gurgling, half-grunting sound as he raked his hands across the jewelry displayed on an old quilt on a stout table.

"More junk made in China," he exclaimed, as he grumbled some more.

As I rushed over to untangle the mass of necklaces and bracelets he had just jumbled, I explained in my best and hopefully, CALMEST storekeeper voice, "No, nothing in this store is made in China or anywhere else outside the United States.... and most is made as locally as possible."

The necklaces he'd carelessly tossed across the table are individually handmade from horseshoe nails by Stephen Wade from Allgood.

The colorful polished rock items are also handmade by a woman patriot, Ruth Woodhall, from Ohio AND Etowah County in Alabama.

But I explained as I almost backed him into a corner, the majority of the items are made by me on this farm from local items grown or produced right here.

No, I don't believe that anytime soon I'll be getting the ingredients for my honeysuckle jelly or my goat milk soap from anywhere any further than my backyard.

"Just thought you were a junk store like everybody else," he said as he left with no apology.

As I sat on my little stool behind my handmade checkout counter, I could only shake my head.

You can preach the benefits of global economy all you want, but in this little homesteader's eyes, we've got to stop before it gets any worse! My disgruntled customer had made a point. Too many stores he'd likely been in were just filled with junk from China! And it's up to us to stop it before it gets any worse!

To me, nothing good can possibly come from growing meat in the United States, having it processed OVERSEAS and then shipping it back to the United States for our families to eat. Somehow the thought of my meal being more well-traveled than I am does not suit me. But it's not just that, we must start examining everything we buy!

A needle breaks in two pieces with one pricking my finger as I'm on the last row of a baby quilt. As I reach to get a new needle, the letters "Made in China" glow gold and mock me from the pink plastic case.

From now on I'm going to be ordering my wooden clothespins from Lehman's Non Electric in Ohio! Their clothespins are handmade by the Amish. The ones that just sprang open on my clothesline letting the sheets drag in the dirt brightly said, "Made in Taiwan."

There's a two-piece screwdriver in the drawer in my office. It's not supposed to be in two pieces, but as I tried to tighten the FIRST screw on a hinged door of my chicken coop, on the FIRST screw, the screwdriver broke nearly in two...not made on this continent.

In my office, there's also an entire drawer of flashlights that will neither flash nor light. All made outside the United States!

And don't even get me started about blue jeans!

If you're like me and you actually live and work in jeans, you better get on the internet and hunt those actually made in the USA or you may get arrested for indecent exposure once you bend over a time or two! My late husband and now, my grown son find that practically every pair of jeans or pants in general will split right in the crotch if they squat down to work.

Is it that they don't know how to sew a seam or is the thread just spider-web thin?

We don't wear that denim just to look like all the wannabe farmers and cowboys...we actually do the work and we expect the jeans to hang in there in one piece right along with us!

An article in USA TODAY notes that every American yearly eats an average of 260 pounds of imported food.

So if we can't trust folks to make a screwdriver that will hold itself together for more than one use or a pair of jeans you can actually bend in, can we really trust what we're eating?

A rash of articles on the internet notes there are simply not enough Department of Agriculture or USDA inspectors to inspect all the food that is imported.

So we go from the simply inconvenient such as a broken needle, to the downright wasteful to the DANGEROUS---such as Chinese drywall (a friend of mine is still struggling to fix all the economic and health issues that caused)---to deadly dog food, to sickening baby formula.

But I see good signs.

Just down my road and on each side of our county from me are young farm families legally selling farm raised chicken and fresh-from-the-farm pork along with farm-fresh eggs and even hickory bark syrup!

Local farmers markets have EXPLODED in the past few years with almost every community having close access to at least one. (There was even a fully stocked farmers market every Tuesday in one of the huge medical buildings in Birmingham where my late husband traveled regularly for treatments!)

And how many folks do you know who have started gardening again? You can know what is in or been spared on your food if you grow it yourself OR if you KNOW the farmer personally displaying it for sale in that wicker basket!

We're done so much to mess up this country and to ourselves with often a generous amount of help from our own government. But all is not lost!

A friend's husband comes from four generations of cabinetmakers. He will custom make cabinets for your kitchen or bathroom for about the same or a little more than you'd pay to get foreign-made cabinets....and the ones he makes won't be made of fiberboard---they will last a lifetime!

Of if you don't know anybody personally, check out the furniture made in many Amish or Mennonite communities around the country!

Some people have asked me why I like their products so much. No, I don't share all their religious beliefs and you have to watch for unscrupulous people even there. But if you find an Amish or Mennonite family who creates with wood rather than builds, you'll likely never look inside a big box store again!

And Amish foods that are Department of Ag approved are the same way---no weird ingredients you can't pronounce.... just old-fashioned goodness!

I hope I educated my recent customer about the Buy Fresh, Buy Local movement before he grumbled out to his car.

We have to check labels. Buy USA made if at all possible. Support your LOCAL farmers and local businesses that sell USA made goods.

It's like that old saying regarding eating an elephant...it may look overwhelming if we look at the overall picture, but even that entire elephant can be eaten if we do it one bite at a time!

Suzy Lowry Geno

The Right to Dry....

(This article is reprinted from my first book because it is the first article of mind that ever won a national award. I was doubly proud because it was displayed throughout that year at the Fess Parker Museum in California!)

I didn't know I was teetering on the verge of becoming a full-fledged criminal.

I didn't know I was part of a national, perhaps worldwide syndicated movement of other such possible law-breaking folks.

I didn't know something I do just about every single day is being considered acts of "civil disobedience" in several parts of the United States.

I didn't realize in some cities or neighborhoods I could face hefty fines and even JAIL TIME if I continued this!

Wow. And I have been doing this for years and years and years simply because I love saving money and love the outdoors.

Who would have thought the simple act of hanging one's laundry on a clothesline would create such havoc across the country?

I can remember as a little girl that practically everybody in the rural South hung their clothes out to dry. We have pretty enjoyable weather here and few folks had clothes dryers ---so why not?

When my first baby was born those many long years ago, my then mobile home had no spot for a dryer. I remember lovingly hanging out her cloth diapers on the short clothesline and standing back proudly to watch them flap in the breeze.

I eventually had more kids and "moved up" to a house AND a dryer, but I still loved hanging the clothes on the line. Can you think of a better experience than climbing into bed after a hard day's work and slipping between sheets fragranced with nothing more than the sun and the wind?

So when our dryer quit more than 20 years ago, I simply didn't replace it. I haven't missed it one bit!

We live about three miles from our area's small county seat and so I'm not governed by too many rules and regulations. Nobody seems to care if I hang out my raggedy towels, worn farming blue jeans and cozy flannel shirts on the lines stretched between metal tee-posts (that my dad had fabricated for my mother more than 50 years ago) in the backyard.

But what do you do when it rains, folks ask? Well, I don't wash on those days. Simple enough.

But if it's just something that HAS to be worn the next day, in the winter it will dry overnight if hung anywhere in the vicinity of our wood-burning heater.

I haven't calculated how much I've saved throughout my lifetime by hanging our clothes out to dry but it has to be in the thousands of dollars, starting with the close to $500 I saved when I didn't replace the blown-up dryer. I've probably spent $10 total on wooden clothespins and maybe another $5 on the wire I sometimes have to re-stretch between the metal posts. But that's it.

Plus there's that added benefit that our clothes seem to last much longer AND I've had the benefit of those sweet-smelling sheets all these years, something no amount of fabric softener can equal!

But now I've discovered that all along I've been part of a bigger movement called "the Right to Dry."

With more folks becoming energy conscious (no matter where you stand on the Global Warming arguments, there's scarcely anybody who doesn't want to save money!), many folks are rebelling against town and city restrictions, covenants in private neighborhoods and rules set out by Homeowner Associations saying you can't even dry your own clothes in your backyard.

There have been "right to dry" initiatives either passed into law or considered in numerous states and some foreign countries including California, Connecticut, North Carolina, New Hampshire, Ontario, Canada, and more.

The North Carolina law, which passed, "invalidates any city or county limitations on energy devices based on the use of renewable resources," namely a clothesline!

The legislation in Florida and Utah prohibits "state or local laws or regulations or private contracts from limiting the ability of dwellers to erect and use clothes lines for the drying of clothes.

According to the Association of Home Appliance Manufacturers, in the mid 2000's there were approximately 88 million clothes dryers in the United States. They said that annually those same dryers consumed about 1079 kilowatt hours of energy PER HOUSEHOLD!

Another national magazine noted that clothes dryers account for approximately six percent of total electricity consumer in the United States.

It doesn't take a rocket scientist to see how hanging clothes on a clothesline will save you money!

But the Christian Science Monitor estimates there are 60 million people living under the rules of 300,000 Homeowner Associations, which mostly bar clotheslines, saying they look unsightly and can decrease property values.

That's not to mention the towns and cities who have similar laws.

That's one reason I live in the country, so I'm not governed by such stupid rules and regulations (and yes, I did use the word STUPID!) If someone wants to live somewhere like a gated community and adhere to all those restrictions, that's their right.

But there seems to be a lot of folks who feel the clothesline ban is not in sync with the renewed call to preserve our natural resources.

At this time there is a national group called Project Laundry List that publishes a regular newsletter and who provides sample letters you can send to your legislators or to your homeowner's group [trying to regain the right of hanging your laundry outside.]

Some from Europe who have posted comments on the Project Laundry List message boards on the internet think it's funny that ANY government or group in the United States even considers banning the hanging of laundry since they said it is common place in most of their areas. They say it just makes common sense to use free solar power to dry the clothes we wear.

I have long wanted to run my home using solar power instead of electricity I have to buy from the "grid" but the cost of solar panels and other equipment has just been waaaay out of my reach thus far. But hanging my clothes on the line is a simple and easy way I can harvest solar energy each and every day for no cost to me whatsoever!

I plan to try to stay out of any national or international controversy.

I also plan to continue to hang my clothes on the line just about every day.

To me it just seems like common sense. A "simple thing" to do to save money and energy.

And if you don't agree, well, I'll think about you when I lay my head on that sweet-smelling, sun-dried pillowcase tonight...

More Simple Times at Old Field Farm

(This article is the only other article in this book that was reprinted from my first book. Many of you know this story because it was published in several national magazines in the 1980s and 1990s. It's not exactly homesteading but it is certainly about family!)

Suzy Lowry Geno

A Chance to Laugh....

I lay shivering on the steel x-ray table.

"Try to stay calm," the radiologist urged. "These tests will be over in a minute" as he stuck an IV into one of my arms.

The next thing I remember is someone shouting "She's coming back! She's coming back!" and awakening to a sickening feeling of bright lights, the hardness and coldness of the steel table, and the feeling I'd been on a faraway journey that left me tired and confused.

The thump thump thump thumping I was hearing as I awakened was actually one of the nurses or technicians doing chest compressions on my heart as Adrenalin was injected to counteract the reaction...

I'd reacted badly to the medications and dyes and evidently almost left this world, my husband, and my two little girls...and would have inadvertently carried the tiny life within me that we didn't even know about at that time.

Roy and I had been hoping for an addition to our family for nearly a year. Beth and Jannea ended their mealtime and bedtime prayers with requests for a baby brother or sister.

Now suddenly my own health problems had forced their way back into the limelight and I underwent four grueling days of intensive tests and x-rays. A routine hospital admissions pregnancy test had revealed the usual disappointing negative.

After scans and tests of practically my entire body, the doctor finally concluded that my pains and severe gastro problems were caused by severe food allergies and he began treatment immediately.

After leaving the hospital, I cheerfully packed my medication and left on a short vacation with my family to the beach.

Since I was not pregnant, I scheduled some extensive dental surgery for the following week.

"Maybe after that I can just concentrate on rebuilding my health and enjoying my family," I thought happily, relieved to be alive and to not have cancer or some other ominous disease.

But five weeks later I was back in the specialist's office, complaining of even more problems. This time the tests showed I was pregnant.

My elation quickly turned to fear as I learned that I had already been pregnant at the time of the intensive tests, numerous x-rays, strong medications, and when I'd been given other medications during and after the dental surgery

After being so careful and "baby conscious" during that year that we had been hoping for a baby, a faulty pregnancy test had caused my unborn child to be exposed to almost every harmful condition from which I had wanted to protect him.

To make matters even worse, I learned that one of my medications had just been added to the Federal Drug Administration's list as one of those that possibly caused birth defects.

"What can we do?" I asked frantically.

My obstetrician was not very reassuring.

He said simply, "If you really want this baby you can only wait and see."

During the next few days, our telephone bill increased dramatically as I called every physician and research specialist about whom I heard or read. Everyone showed concern but no one could give me the reassurance I sought.

This was before the time of ultrasounds that show tiny babies now sucking their even tinier thumbs at just a few weeks of age! And the only thing an ultrasound would have done back then would have been to have reassured me that the baby was fine or given me time to prepared for whatever problem awaited his or her birth.

This was also years before everyone's quick access to the Internet so I visited our library's reference section and began reading every book and magazine article that deal with the many causes of birth defects.

I was strengthened and awed by the courage and patience that many parents of special children possessed when they faced the staggering responsibilities of a handicapped or gain damaged child.

But in examining newspaper and magazine articles on genetic research, I saw the word abortion often listed as a "simple and convenient alternative" to that of bringing an "imperfect child" into this world.

Back at home, I found the small booklet my doctor had given to me before the birth of our oldest daughter. I quickly turned to the illustrations of a baby inside its mother's womb.

The brightly colored pictures illustrated the fact that by the time we had discovered I was expecting, my baby already possessed tiny arms and legs, and its small heart was beating steadily.

From the moment of conception, this little creature had grown from a minuscule part of my husband and me and had begun its own individual and unique existence.

During the next few months as I felt our baby kick and move, I would not help wondering if his limbs would be shortened or otherwise deformed. Could our human errors have interfered with his brain development?

My prayers went up constantly for this new life that was surrounded by so much love.

I was certainly not being brave or heroic---I was fighting a daily battle with my emotions.

I was uncertain about so many things. I only knew that whether our baby was perfectly healthy or whether he possessed formidable handicaps, he would still be our gift from God and would have his own special purpose in life.

I reread Dale Evans Rogers' book Angel Unaware, about her handicapped child, Robyn, and I prayed that God would give us the same kind of faith in whatever situation we faced...

Many months later, I visited our local Wal-Mart to have photographs taken of the perfectly healthy son with whom God had blessed our family.

As I was writing a check while waiting for the photographer, another woman waiting with a friend nearby, asked my son's age. She saw my obstetrician's business card sticking out of my checkbook.

"I was pregnant but I didn't know it,
she said.

She then explained that she had the identical tests I had undergone with many of the same medications.

"Later my doctor told me that if there was a defect, it probably couldn't be determined by amniocentesis, but he thought there was a big chance of fetal deformity. I decided not to have the baby," the woman went on...."I can always have other children later." Her voice trailed off....

Watching a few minutes later as my laughing, handsome little boy looked up at the camera, tears stung my eyes.

My heart was aching for the other woman's little child who never had a chance to laugh....

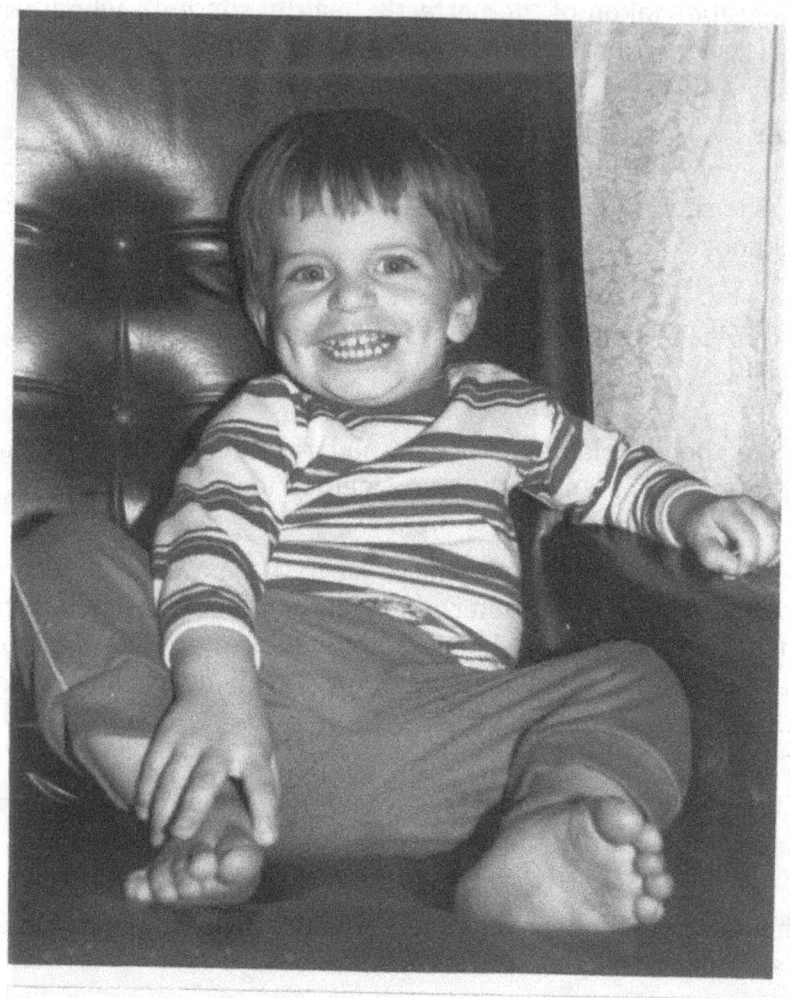

Nathan shortly after the original article was written.

Suzy Lowry Geno

Finding God in the Barnyard

You see it all the time on social media.

Someone's kids had an awful wreck totaling their car, but walked away without a scratch. The overjoyed mama posts: "God is good all the time."

Someone else survives a massive heart attack and comes home from the hospital and is able to simply resume his normal life. He joyfully writes, "God is so good!"

There's a big financial gain where there was once desolation, happiness follows months of sadness, joy springs forth.

There are slogans and memos on Facebook, encouraging words printed on tee shirts, even special vinyl letters cut out to go on your vehicle's back windshield.

And don't get me wrong!

God has blessed me and this little farm exceedingly abundantly more than I would ever deserve! And I thank Him and praise Him daily for all those rich blessings!

But so many times when I see a Facebook post or even a Tweet, it makes me stop and think.

If the kids in the wreck were badly injured or even killed, does that make God bad?

If the friend or neighbor with the massive heart attack never even makes it to the ER before passing away, does that make God bad too?

And what if you didn't get that raise or your farm is foreclosed on no matter how hard you've worked and what you've tried to do, does that put God in the negative???

I'm no great theologian. I'm just a gray haired, pudgy homesteader. But I do think too many times we are putting limits on God.

I wrote four years ago about how my heart changed as I was driving across the viaduct in Birmingham just before my husband Roy and I reached Kirklin Clinic and the nearby UAB Hospital.

We were going for an all-important PET scan to see if the months of chemotherapy and weeks of radiation had killed the cancer in his body.

I'd made the comment that if that test came back clear you might see me on the TV news that night from Birmingham, leaping and running across the sidewalks, praising God like an old-time country preacher!

But as we grew nearer our destination and nearer time for that all-important test, my heart was greatly conflicted.

There was no doubt in my mind that would have indeed been shouting from the mountaintops if that test had come back showing the cancer was gone.

But the simple message I realized as I drove on toward that sandy-colored brick building was that I should be willing to praise God ALL the time, no matter what the tests results were because we had already been blessed beyond measure in our lives.

I didn't say anything to Roy, but I was thoroughly chastised. And I haven't forgotten that lesson.

It's a religious debate that goes back as far as Job in older times. Why do good people suffer? Why do so many bad things happen in life? Why are some people miraculously healed when others who appear to have even more faith are allowed to suffer and die?

I guess living on a farm or homestead makes a lot of folks closer to life and death than if you simple live in the suburbs.

It seems, so far, most of 2016 has been a learning year for me. For somebody who doesn't like to go to doctors because they'll find something wrong with you and who puts up with lots of pain from a torn hamstring to a badly bruised hip without pain medications at home because of my allergies, I guess I was getting a little complacent.

But when I awoke in the early morning hours of December 29, 2015, I was ready to see a doctor---ANY doctor---because I was in such pain.

A visit to the ER was followed by seeing a new specialist to treat kidney stones and other problems. And all seemed well again...then came March.

My youngest daughter and I got some sort of flu that lasted and lasted and lasted...for more than two weeks I coughed until it seemed every part of my body was sore. There was no energy in spite of shots of steroids and more. I hadn't felt this badly since I was a little girl and suffered through the old-time flu, measles, and chickenpox!

Then there was another early morning trip to the ER, an ambulance ride to an out-of-county hospital, and emergency surgery to remove a CLUMP of kidney stones! Yipes!

And while all this was going on, my little homestead kept right on moving forward as well.

Everyone knows my goats are my babies, no matter how old they are. Johnny-goat's twin sister died the same week Roy died, nearly four years ago.

But Johnny-goat hung on until I was suffering from the flu. I prayed he wouldn't make it through the night because I didn't want to see him suffer. I'd already made up my mind to call the vet in the morning if he was still with us.

He mercifully did pass away. But my daughter and I both were barely getting basic essential chores done because we were both so sick. How would we bury a goat (a Nubian-Boar cross) that was the size of a horse?

A neighbor came to the rescue with a backhoe-looking-contraption, easing that worry form my mind.

Then it seemed a whole passel of the guineas were determined to commit suicide by standing in the highway! My sweet little birds that we'd hatched in the incubator last year, that I'd raised in a pasteboard box with a heat lamp, and that I loved for their wacky lifestyle AND the fact they were keeping the snakes away!

During one day at the height of my flu I had to walk out into the highway and retrieve THREE guineas who met their demise at three different times!

I did not miss a single day tending to my animals during the entire five to seven years my husband was sick. I often did chores by flashlight in the mornings and then again by flashlight at night when we'd come back from his treatments.

But as I was in the throes of the flu AND THEN had to have emergency surgery for the kidney stones, I realized things were completely out of my hands. (Especially after the original ambulance ride, folks thought I might even be having a heart attack myself!)

But you know what?

If I'd had a heart attack and never made it to the hospital, God would have still be good!

Even when I was truly suffering pain that was pretty intense, God was still good!

You can argue theologies, religion and doctrines from now until doomsday, but you'll never ever find all the answers. I don't care if you read all the holy books of the world in their original languages and have diplomas covering your walls.

Some would even argue that all the trials, tribulations, and tragedies of this world prove there is no God.

But even though this simple homesteader does not have all the answers, I do know this.

This week, a small game/Easter egger hen hatched six chicks in our dog, Maggie's, pen. Those little bright balls of fluff don't know anything but that their mama is there taking care of them.

There's a Scripture about God spreading His love out and sheltering us as a hen protects her babies.

I don't have all the answers. I don't even understand all the questions. But this I know: whether I am here on this Earth or gone on to the next, I am sheltered and protected with God's love just like those tiny chicks.

We may walk through the valley of the shadow of death, but, in the long run, it will not matter because traveling through the valley is the best way to reach the mountaintop!

God IS good, ALL the time!

Suzy Lowry Geno

Johnny-goat

Work with your hands

"You have Granny's hands." With that simple statement my niece Jeanne recently started me on a whole new way of thinking.

I don't think there was anything really unusual about my mama's hands. My first memory of mama's hands was the extremely wide shiny gold wedding ring she wore on her left hand. No dainty little diamond for her, but a substantial heavy but pretty band that represented her solid marriage to my daddy.

I can't really recall her ever polishing her fingernails, although she probably did that as a teen growing up with a houseful of brothers and sisters.

But in my mind mama's hands were always busy. They were often red and chapped when I was little from washing and bleaching clothes and running them through the old wringer washer and then hanging them to flop on the clothes line until dry.

I was a sickly little kid and some of my sweetest memories are of mama rubbing salve on my chest or in the "hollow hole" of my throat to help me breath, and often her bare work hands almost felt scratchy against my tender skin. But oh what a comforting feeling, the scratchiness didn't matter one bit!

She wasn't one to do a lot of "busy" work like my Great Aunt Cora Lowry who was always crocheting or even tatting with her fingers flying.

But mama did teach me how to thread a needle and how to make the basic embroider stitches. Then her hands guided mine as I turned the wheel on the old treadle sewing machine when my feet would barely reach the treadle, until I eventually made almost all my own growing-up-year-clothes.

I can remember mama helping my Aunt Lucille Evans in her florist shop, busily twisting green florist tape around the stems of flowers as they fashioned wreathes and "sprays" during busy decoration days.

Later as daddy eventually owned his own construction company, mama's hands were able to rest a little more, although she never ever owned a dishwasher!

She spent many enjoyable years learning to china paint from the late Wileana Bynum Buckner and also painted all sorts of beautiful or funny scenes on the woodworking crafts my daddy built during his retirement. A flick of her wrist and a dappled pony pranced across a child's chair. A careful movement of her fingers and the brush filled in a flowered swag drifting across a bread box for somebody's country kitchen.

And lastly I think of her hands, holding a stubby pencil as she filled in find-a-word puzzles or crossword puzzles in little books to fill up her long widowed days...writing simple grocery lists for me to fill for her...every now and then I still find one although she's been gone nearly a decade!

So my hands are my mama's hands. Nothing fancy. Kind of short stubby fingers.

But unlike mama, I kept the nasty habit of biting my fingernails until I was in my mid-thirties so I surely had nothing to paint or polish! (I tried everything to quit biting them---bitter stuff that you painted on the ends of your fingers and even taping each of my nails! I only quit biting them when I was confined to the hospital for ten long days in my mid 30s with three IV bottles in each arm! I don't recommend that as a nail biting remedy but it worked!)

So my hands have just always been there when I needed them.

Daddy's hands were different. At least two of his fingers were considerably shorter than the others from accidents with saws through his years of carpentry and the late Dr. J.L. Wittmeier literally put daddy's right had back together with 543 stitches during several hours times after a terrible accident with a table saw! There was no fancy physical therapy then. The doctor sewed, they prayed, and daddy squeezed a small red ball for therapy until he regained most of the use of his always-stiffened hand.

Then I think of the hands of my piano instructor at Hillsborough Community College in Tampa Florida. Dr Patricia Trice's long elegant fingers were so born to play (the exact opposite of my shorter stubby ones!) that the college brochure featured a wide circle just showing her hands gracefully playing a keyboard...

My late cousin Jack Lowry told about laying in front of a fireplace while his Grandma Lowry (my great-grandma Lowry) spun yarn on a spinning wheel, her fingers seeming to dance in the shadows of the firelight, as he was not yet three years old.

Just in the last month, I've been privileged to be sent a photo of my other Great-Granny, Sally Roach Smith, and oh my goodness, I look so much like her it is amazing. I've strained and strained trying to see her fingers holding her little daughter, my great aunt...

So I've carried these two hands and ten fingers around for more than 64 years now, just expecting them to react and behave like I want them to. Giving them little thought unless I stuck a splinter into one, ashed one putting wood into the heater, or jabbed the needle too far as I was quilting!

I didn't realize until this morning that NAILS magazine was showing their industry showing record-breaking growth of 7.47 billion dollars for 2012-2013 (and I didn't even realize they had a magazine!)

And while lots of kids are really getting into robotics now, I've just discovered that our FINGERS are REMOTE CONTROLED!

According to E-Hand.com "Of course in one sense, we work all our moving parts by remote control...the control center is your brain." But it goes on to say that our fingers are special because "there are no muscles inside the fingers!" "The muscles which bend the finer joints are located in the palm and up in the mid forearm, and are connected to the finger bones by tendons, which pull on and move the fingers like strings on a marionette."

So these stubby little fingers with which I type, quilt, sew, play the piano, and do all manner of things, are not only just like my mama's (and possibly two sets of Grannies as well) they are attached to two hands that are just a part of God's miracles.

In Proverbs, a Godly woman is described and part of her work is "she puts her hands to the distaff, and her hands hold the spindle," (Proverbs 31:19)

But I think my very favorite verse and one that I'm struggling to live by (complete with hands that look just like my mama's!) "Make it your ambition to lead a quiet life, to mind your own business and to work with your hands, just as we told you, so that your daily life may win the respect of outsiders and so that you will not be dependent on anybody." (1 Thessalonians 4:11-12)

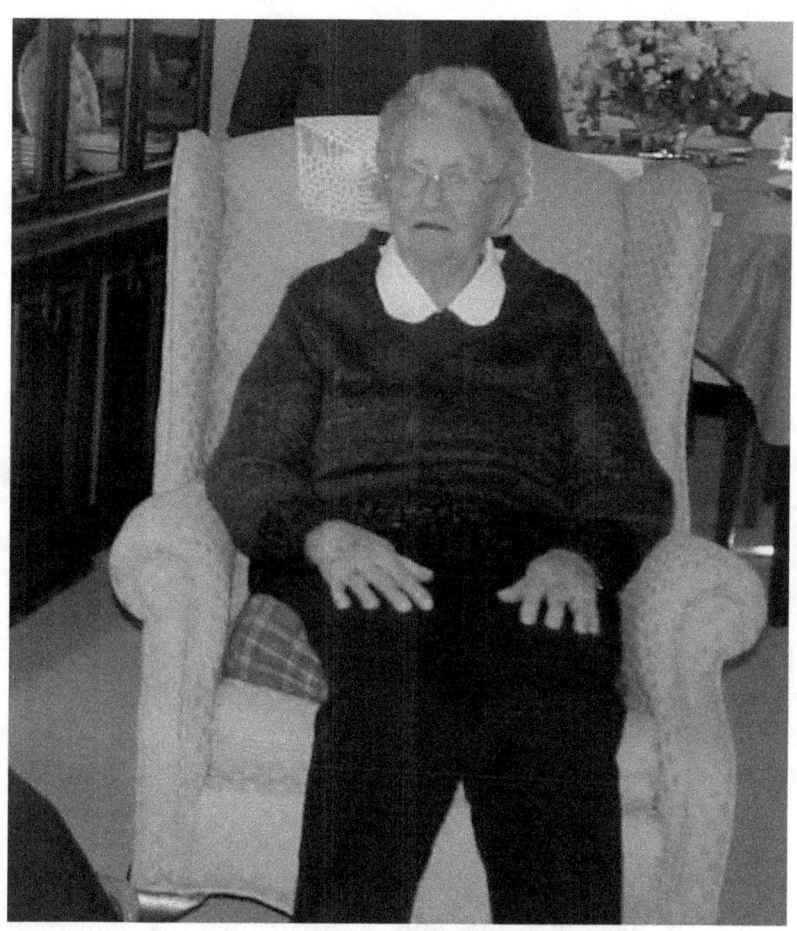

My mama, Inez Inmon Lowry

Not the lone ranger

I've always said that my ideal situation would be a cabin in the middle of 100 wooded acres with a big, TALL fence encircling all around!

Who hasn't just wanted to "get away from it all" and live out their lives unencumbered and unbothered?

I've always been a really shy person. I've told before how shocked my mama was when I announced that I was working as a newspaper reporter/photographer about 35 years ago. She couldn't imagine me going out and interviewing ANYBODY, not less some of the "important" people I was required to meet with every day.

But I found that my notepad and my camera were good crutches and as long as I held them tightly in my hands I could interview anybody from the hottest country music stars right up to the vice president. But inside I was still that shy little girl who would rather retreat to her world of books and fantasy than deal with most people!

While my husband was so sick those several years, it was almost like it was just me and him against the world at times....we'd fight his illness with the latest technologies and medicines in the big city, but retreat to our little country home where we could just cocoon and be left alone!

Then after his death I became even more of a hermit. And that's not all bad!

I am so blessed that I am content and at peace in my aloneness! Every time I see somebody post on Facebook that they are bored, I am just amazed! I have enough projects and things I'd like to do on my little homestead that I could stay busy 24 hours a day if I could just stay awake!

I guess those first few months were times of healing for me and I reveled in being alone, sitting by the wood heater quilting, knitting or crocheting with a purring cat at my feet.

I wasn't completely a hermit in that folks came to the tiny general store on my farm and I always enjoyed visiting with them. But I pretty much stayed on these 15 acres.

Some relatives suggested I go on a cruise...oh my goodness! To me, a cruise sounds like pure hell on earth! Small cabins, confined spaces, no farm animals...certainly not my cup of tea.

Then a group of "girls" from my high school class decided to go the Smokies! I loved reading their antics and seeing their photos on Facebook every day, but I couldn't imagine me up there with them shopping, sight-seeing, and traveling around.

My mama used to laugh and say it was just the Lowry in me! And it did seem I took after some of my relatives who preferred staying at home on their farms!

But after those long months of dealing with Roy's illness, while I was content being alone and staying on my little farm, something was missing.

I could worship my gracious and great God right here on the farm; that was true. And I was able to attend church with some very much beloved folks every Sunday via the Internet. (Much of while Roy was sick we were unable to attend a real church service because of the close proximity and the possibility of his exposure to germs)

But I ventured out one cold, Sunday morning and crowded into a little rented country church jam-packed with families. And a little fire began to glow in my soul again.

A few months later, that church was able to buy a permanent home only four miles from my homestead!

I was there that first Sunday in the new building and I've been there ever since.

It's not big or fancy. There aren't all sorts of fancy-sounding programs. There's just Christ and Christ alone, and oh how that warms my soul...

But what was going on?

Bro. Cliff Cook, one of the two pastors there, summed it up during two different sermons, one just last week. "We aren't meant to be Long Rangers!" he emphasized again.

You probably remember the "long Ranger" on TV from 1949-1957 and available even now in reruns. (NO, I'm NOT old enough to remember it on radio!)

Clayton Moore played the Lone Ranger who was the only Texas Ranger not killed in a terrible attack. He donned a mask and spent the rest of his life (or so the story went) riding his horse, Silver, on solitary missions to help others. But even he was aided by his sidekick Tonto.

But Bro. Cliff made his point to me with just a few statements.

God calls us to be witnesses to others and while I can do that through my writings, sometimes that means getting out of our comfort zones and going where those others are.

But most importantly, Bro. Cliff explained that while we are bearing fruit we NEED the fellowship of other believers to experience a true relationship with Christ.

While even Christ needed His solitude and often was portrayed going into a garden or some other place alone to meditate and pray, he had His chosen 1 often around Him AND he was always traveling to and fro where there were people in need.

Our other pastor, Brother Eric Hixon, often encourages those in our church to "love on" one another, not just during times of peril or tragedy but through the everydayness of life.

We've seen that so many times in the past, but we really saw it illustrated close to home two weeks ago.

I had a little health scare and my youngest daughter, Jannea, was awakened from a sound sleep to take me to the emergency room. She texted just one message to Bro. Eric and, before you know it, there were folks praying for both of us throughout our county. She received texts and emails throughout that day asking if we needed any help and letting us know they were praying.

I was able that night to hobble around and tend to all my animal chores---shutting chicken doors, feeding goats and making sure rabbits were OK. But the love shown throughout that day was special...and it happens every day in some form or fashion to other folks as well.

So while you won't find this simple, gray-aired goat-woman hopping on a cruise ship any time in the future or enjoying the music row in Branson, MO. There is a special fellowship that has called me home to another special peace...

Suzy Lowry Geno

Thanks

Suzy would like to especially thank the folks at the Alabama Cooperative Farming News for their permission to reprint these articles in this, her second book. Every article in this book, with the exception of "A Chance to Laugh" was first printed in the monthly Co-op magazine!!!

About the Author

Suzy Lowry Geno is a homesteader who loves her family and her animals, runs tiny Old Field Farm General Store on her property, and strives to live a peace-filled and simply-lived life.

Before becoming a full time homesteader, Suzy was an investigative reporter/photographer for more than 30 years for daily and weekly newspapers in north central Alabama.

She has received numerous state and national awards for her writings including the national awards mentioned throughout this book for her regular column, "Simple Times" in the Alabama Cooperative Farming News, was chosen the "Media Person of the Year" for back-to-back years 2002 and 2003 by the Alabama Farm Bureau for her coverage of agricultural issues and interests, and in the past received the Alabama Press Association's awards for Best In-Depth Series, Best Community Service, and third place feature photo for the state.

www.ingramcontent.com/pod-product-compliance
Lightning Source LLC
Chambersburg PA
CBHW071608170426
43196CB00034B/2224